BIOLOGY

Helen Nowacka, B.Sc.

**Published by Intercontinental Book Productions in
conjunction with Seymour Press Ltd.
Distributed by Seymour Press Ltd., 334 Brixton Road, London, S.W.9 7AG**

1st Edition, 9th Impression © Intercontinental Book Productions 1974

Printed in Great Britain by Unwin Brothers Limited
The Gresham Press, Old Woking, Surrey
A member of the Staples Printing Group

ISBN 0 85047 211 3

11/77/12

Contents

How to Use Course Companions

Course Companions have been prepared by teachers who have detailed and first-hand experience of examinations and examining. They direct your attention to those areas of the subject which matter most, and which frequently cause confusion and difficulties to students.

Studying involves far more than merely sitting down and reading a book. Real study calls for much more active participation on the part of the student. Research shows that regular study and frequent revision are of far greater value in learning than last minute cramming. It is essential to work steadily throughout your course, making sure that you understand each topic before moving on to the next. To get the most out of these books you should first read that part which is relevant to the particular topic you are studying. Making your own notes and diagrams can be of considerable assistance in re-inforcing what you read—the mere fact of writing being a tremendous aid to memory. Working through the printed examples, stage by stage, and trying other examples, perhaps from past examination papers, will also be extremely useful.

Concentrated attention is physically very tiring, and can therefore only be maintained for short periods of time. By emphasising vital parts of the syllabus, **Course Companions** make study time more effective.

General Advice

There is no short cut to examination success. If you have done no work, you will not pass, and no advice will help you. Many candidates decide that they will do just enough work to scrape through. This is a very risky attitude, far too much is left to chance. You should aim at a good pass, the margin of safety is wider and you will learn more about the subject, which is at least as important as passing the exam. Also, examining boards tend to set more questions which depend on the understanding of biological principles rather than on factual knowledge.

REVISION

Obtain previous examination papers and a copy of the syllabus. Plan a revision timetable based on the syllabus. Allow yourself enough time to revise all of it, but do not start too early. Allow about 8 weeks revision time, to be completed at least a week before the examinations start. Do not study for very long uninterrupted periods, divide your time into hourly sessions. As part of your revision, at the end of each section, select relevant questions and answer them in the time allowed in the exam. By studying the syllabus and previous examination papers you will notice that there are sections on which questions are always set. If you have insufficient time to revise the whole syllabus obviously these sections must not be left out. You must revise the whole of the section, if one part is missed out and the examination paper concentrates on this topic, the work you have done on the rest of the section will be of no use. Some examining boards set a compulsory section consisting of a number of short questions. In order to complete this section you must study the whole syllabus.

THE WRITTEN PAPER

Read the instructions and all questions carefully. Eliminate those questions you are unable to answer. Read the remaining questions and select the stipulated number. You must attempt the correct number of questions. If you answer only four out of five questions, it is almost impossible to make up the marks you could have been awarded for your fifth question. Where you have a choice of questions equal marks are allocated to each answer, therefore you should spend the same amount of time on each of the questions. Candidates often devote most of their time to those questions they feel they can do well, sometimes only two out of five questions. This decreases the total marks for the examination.

HINTS ON ANSWERING QUESTIONS

Read the question very carefully before you write anything, make sure that you understand exactly what the examiner means. Make a rough plan of your answer before writing it out in full. Where a question is divided into parts it may be unnecessary to plan it further. It is particularly important to map out the answers to questions which begin with "discuss", "compare" or "contrast". A plan will help you to write concisely, and to answer in a logical sequence and lessens the chances of missing out important points. When revising, practice planning your answers in the way shown below.

Example:

Discuss the factors which affect the transpiration rate of plants.

1. Mechanism by which plants transpire.

2. External factors
   (a) relative humidity
   (b) wind
   (c) temperature
   (d) light

3. Internal factors
   (a) condition of plant
   (b) biological clocks

Write concisely and keep to the point. This saves time and no extra marks are awarded for sheer quantity or for irrelevant detail.

FORM OF QUESTIONS

The following outlines some of the ways in which biological questions are often phrased and how they should be interpreted.

Tabulate

A table consisting of two or more columns should be drawn up. The columns should show any similar or differing features side by side.
Example: Tabulate the main differences between wind and insect pollinated flowers.

Wind pollinated flowers	Insect pollinated flowers
Small inconspicuous	Large brightly coloured
No nectaries	Nectaries
Large pendulous anthers	Smaller enclosed anthers
Large quantities of pollen	Smaller quantities of pollen
Smooth light pollen grains	Rough coated or sticky pollen
Protruding large feathery stigmas	Compact enclosed sticky stigmas

List

Simply write the relevant information briefly, numbering each fact, 1. . . ., 2. . . ., 3. . . . and so on. Questions which ask you to list the similarities and differences are best answered by a table consisting of two columns of facts. It is not really correct to list or tabulate your answer if you are asked to give an account.

Outline

The examiner wants the main facts only, detail is not required.

Discuss

Particular care needs to be taken with questions of this type. Candidates frequently write a lot, most of which is irrelevant or mere padding. It is particularly important that a rough plan is made out before the question is written in full. Sometimes you will discover before losing too much time, that in fact you have very little to write. Do not attempt this kind of question unless you understand the topic really well.

Compare

Discuss similarities and differences between the subjects which are given in the question.
Example: Compare the structure and functions of xylem and phloem.
Avoid writing a list of characteristics of one tissue to be followed by a similar list for the other. Questions of this type are best answered by selecting such features as transport and strengthening qualities, and discussing how they apply to phloem and xylem. Labelled diagrams of both tissues should be drawn side by side so that structure can be compared.

Contrast

The same approach as in the previous example, but the emphasis should be on the differences between subjects.

Compare and contrast

All similarities between subjects are discussed first, followed by a similar discussion on differences.

Distinguish between . . .

This means to show the differences between two or more subjects and should be dealt with in a similar way to "contrast" type questions.

An illustrated account

Diagrams should be given, followed by an account stressing the functions of the parts labelled.

Annotated diagrams

These are labelled diagrams. In addition, notes are given under the labels, these should be brief and in answer to the question. Unless the question specifically demands it, a further written account is unnecessary.

Diagrams

Instructions at the beginning of examination papers tell the candidates that credit will be given for labelled diagrams where relevant. Some questions ask for diagrams, so you must practice drawing as part of your revision. Where diagrams are not specifically asked for, you should include them if they clarify the answer, or reduce your written description. Avoid duplicating information. No extra marks will be awarded for writing what has already been given in the diagram. Duplication also wastes time.

Drawing: Use a sharp H.B. pencil. In addition you will need a soft rubber, a ruler and spare, previously sharpened pencils. It is really amazing how many candidates waste time sharpening pencils and borrowing rubbers during the examination.

Diagram outlines should be bold and continuous, do not sketch. Diagrams should be large. Some examining boards issue answer booklets with spaces for diagrams; these indicate the size required. If you find diagrams difficult, you may find it will help if you faintly draw the diagram first, then make over it a single decisive outline. Only use coloured pencils where they clarify a diagram or save time. Avoid shading for purely artistic reasons, and drawing many identical structures. For example, it is a waste of time drawing every scale on a fish, or every cell in a leaf. Draw carefully a few representative structures only.

Labelling: Every illustration should have a title. Diagrams must always be clearly labelled, preferably in ink. Draw all the guide lines using a pencil and ruler. Try to arrange these so that the labels will be horizontal and fairly evenly spaced around the diagrams. Guide lines should never cross. If your diagram involves using colours or shading a key must be included, if the labels have not explained the significance of the colours.

Questions on experiments

Diagrams of apparatus should be drawn and labelled, and any special features indicated. Do not give a written description if this duplicates the diagram. Explain experimental procedure where necessary. Wherever possible, an experiment must have a control. This is an additional experiment differing from the main one by the one factor being investigated. The control will show that no other factors are responsible for the result.

THE PRACTICAL EXAMINATION

Some examining boards include a practical examination as part of the paper. Therefore living material should also have been studied using a hand lens, and where necessary a compound microscope. Simple experiments should also have been carried out by candidates. As part of the exam you will be expected to make careful observations, recording them in writing and by means of accurate, labelled drawings. Some boards will provide photographs and diagrams instead of specimens, and questions are set on these. Although not strictly a practical exam, in order to do such questions, experience in practical work is essential. When revising theory always study the relevant practical work with it; never revise practical work on its own. Timing your answer should have the same consideration as in the theory examination. Sometimes the number of marks allotted to each question is given; this indicates the amount of time you ought to spend on each question.

The following is a list of practical work which could be set in a practical examination.

1. Drawing and labelling the external features of insects and other small invertebrates.

2. Drawing and labelling parts of the skeleton.

3. Making written observations on teeth.

4. Studying the behaviour of small invertebrates, particularly movement and feeding of aquatic animals.

5. Making labelled drawings of external and internal structure of flowers, fruits and seeds.

6. Drawing and labelling parts of plants, e.g. bulbs, corms, rhizomes, tendrils, twigs, etc.

7. Making observations and comments on simple experiments.

8. Performing food tests.

Variety of Living Things

Questions on the variety of living things are of the following types:
1. Examples of plants and animals are given, you are asked to show why some are plants and others are animals.
2. Certain processes are given, for example respiration, reproduction, nutrition, etc., and candidates are asked to compare one or more of these in a wide range of organisms.
3. Questions ask for comparisons of body or egg structure in a number of organisms.
4. Candidates may be asked to describe and compare the life histories of organisms. A knowledge of how certain organisms such as earthworm affect man may also be expected.

General questions

1. What are the characteristics that lead you to classify an organism as an animal or a plant? Using these, classify Amoeba, Spirogyra, Mucor, a mosquito and a buttercup.
2. Describe with the aid of diagrams the general structure of Mucor and main features of its life history. How does the nutrition of Mucor, which is a plant, resemble that of animals?
3. Compare the eggs laid by a frog with those of a bird. How do you account for the fact that mammalian eggs are minute, even though mammals, in general, are much larger than birds?
4. State the difference in external structure between (a) a mammal and a bird, (b) a month-old tadpole and a fish, (c) an Amoeba and Hydra.
5. Draw a large labelled diagram of a named fish. Describe how the body of a fish is adapted to life in water. How does the fish feed and breathe?

Framework answers

1. Basic difference is one of nutrition. Animals take in and digest organic food in cavity in body. Most plants synthesise organic foods from simple inorganic substances. Do so by means of chlorophyll. Animals move about to get food, plants do not. Plant cells encased in cellulose, this absent from animal cells. Animals grow to a maximum size; plants grow throughout life, mainly terminally.

Body of animal compact, that of plant is branched. On these differences can classify Amoeba, mosquito as animals, Spirogyra, buttercup as plants. Mucor as a plant; non-motile, branched terminal growth, non-living wall, but no chlorophyll, absorbs food through surface.

2. See opposite page for diagrams.

Body or mycelium of fungus is mass of threads, hyphae. These are hollow tubes with no cross walls. Wall of hypha does not contain cellulose, is lined by layer of cytoplasm in which are embedded many nuclei. There is large central vacuole. After mycelium established, some hyphae grow erect, ends swell to form sporangium. Within, protoplasm forms uninucleated spores each with protective wall. Sporangium opens to release spores, these dispersed by air and insects. Spore wall breaks open when on suitable food, protoplasm grows into new hypha, this later grows into new mycelium.

Sexual reproduction only occurs when two forms of mucor (+ —) grow together. Nuclei from the two hyphae are brought together as shown in diagram, and fuse in pairs. Resistant zygospore formed, is dispersed, germinates after dormant period. One hypha grows to form sporangium which releases spores, these dispersed, germinate to form new mycelia.

Mucor cannot synthesise food. Produces enzymes which digest organic material. Soluble food absorbed.

3. Frog's egg unfertilised, bird's egg fertilized internally. Frog's egg black with small whitish area on lower side. Has moderate amount of yolk. Bird's egg yellow, large amount yolk, small clear germinal disc on top, development already started during passage down oviduct. Therefore few cells of embryo already present. The yolk is suspended in a large amount of watery albumen. Frog albumen layer thin and dense at first, swells rapidly on contact with water. Bird's egg has two shell membranes enclosing air chamber at one end, whole surrounded by protective, porous calcareous shell.

Mammalian egg minute because of small amount of yolk stored. Only sufficient food reserve required for very earliest stages of development. As developing embryo passes down fallopian tube to uterus it can obtain some food from the surrounding tissue fluid. When in uterus it becomes implanted in uterine wall, absorbs food from this and later, with formation of the placenta, gets all

Mucor

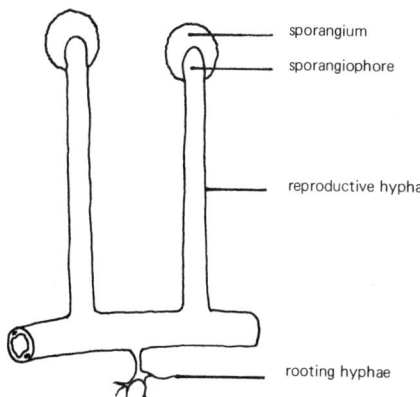

sporangium

sporangiophore

reproductive hypha

rooting hyphae

Sexual reproduction of Mucor

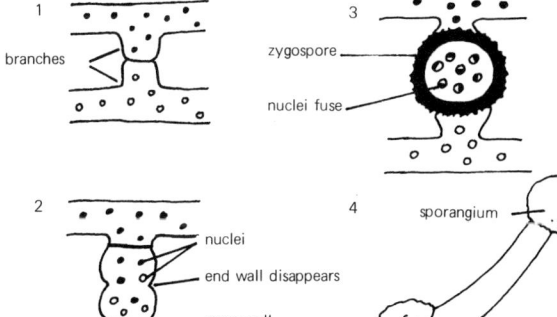

1

branches

2

nuclei

end wall disappears

cross wall

3

zygospore

nuclei fuse

4

sporangium

13

necessary supplies direct from mother by diffusion from her blood to its own.

A bird's egg must contain sufficient food and water for development of chick to be completed before hatching. The "white" is essentially water storage material, hence its bulk. Some water is produced during respiration. Although some water may evaporate through the shell, this is minimised by the covering body of the incubating parent.

4. (a) Present in mammal such as rabbit; sweat glands, fleshy lips with teeth in jaws, external ears, all limbs for locomotion on land, separate anus, urinogenital openings, mammary glands, scrotal sacs in male. In a bird such as the robin, there are feathers, scales on legs, no sweat glands, horny toothless beak, no external ears, anus and urinogenital openings in cloaca, no mammary glands or scrotal sacs, very short tail with long feathers, preen gland. Forelimbs modified as wings.

(b) Tadpole has smooth skin, no scales, internal gills are covered by skin-like operculum. Small aperture, spiracle, on left side only. Tail surrounded by transparent fin. Fish; skin covered with overlapping scales, operculum on each side. This is large bony scale covering slit communicating with the gill cavity. Unpaired dorsal and ventral fins as well as tail fin. Paired pectoral and pelvic fins. All fins supported by fin rays.

(c) Amoeba has a minute irregular shaped body when active, but is rounded at rest. Continuous surface of cytoplasm, jelly-like. Can extend pseudopodia from any part of surface, there is no mouth. Hydra has cylindrical body, mouth at one end on oral cone, is surrounded by 8–12 tentacles. Projecting from the body surface of cells, are cnidocils of nematoblasts (or stinging cells) mainly on tentacles, and the processes of the sensory cells.

5. See opposite page for diagram.

Stickleback: Water gives greater resistance to a moving body than air, therefore body of fish stream-lined. It is smooth due to backwardly overlapping scales. These camouflage body, dark grey dorsally, prevents detection from above water, it is silver ventrally, protects from attack below. Scales waterproof, prevent too much water uptake by osmosis. Tail and flexible body push on water

Stickleback

for movement. Other fins stabilise and direct body. Swim bladder prevents sinking when fish stops swimming. Fish has gills for oxygen uptake in water. Lateral line is sense organ to detect movement in water. Eyes adapted to seeing in water, to judging position of objects out of the water. Nostrils lead to organs of smell only. These detect food in water.

Feeding: Carnivorous, feeds on small pond animals or carrion. Food detected through organs of smell, is sucked into mouth, usually swallowed whole, but teeth can grasp food.

Exchange of gases: Mouth open, floor lowered, operculum (gill cover) closed. Mouth closed, floor of mouth raised, water is forced over gills and out through opening under operculum. Oxygen from the water absorbed into blood in gills, carbon dioxide passes into water from the blood.

Stickleback

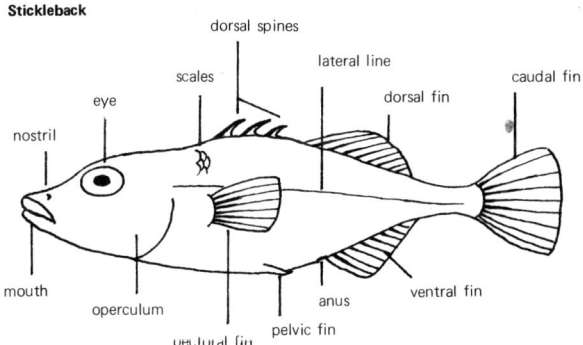

Cells

Units of protoplasm, differentiated into cytoplasm and nucleus. Nucleus contains chromosomes which control cell by passing instructions to cytoplasm. Here enzymes do work to carry out respiration, nutrition and other processes necessary for life. In cytoplasm are number of minute bodies, including mitochondria. Enzymes bringing about respiration are concentrated in them. Surrounding the whole cell, nucleus and vacuoles, are membranes. These control movement of substances in the cell, also passage of materials into and out of the cell.

In multicellular organisms there are many different types of cell groups (tissues). These all originate from the single fertilised egg cell. How these different tissues arise is not understood.

General question

Describe the basic structure of an animal cell. How do the following differ from this; a nerve cell (neurone), a red blood cell and a cell from the cortex of a plant stem?

Framework answer

Labelled diagrams of the cells mentioned should be included. The difference between the animal cell and one from the cortex of a plant stem will show the basic difference between plant and animal cells.

Animal cell: Compact mass of protoplasm, central nucleus with chromosomes in nuclear sap, surrounded by nuclear membrane. In cytoplasm, centrosome near nucleus, chondriosomes. Vacuoles, if present, minute. Plasma membrane is the outermost limiting layer.

Neurone: Not compact, fine cytoplasmic fibres extend from cell body, long ones usually enclosed in myelin fatty sheath, the fibres branched at ends.

R.b.c.: Biconcave disc, no nucleus, haemoglobin present. Cortex cell: Much larger, cellulose wall around the plasma membrane. Large central vacuole, leucoplasts in cytoplasm, chloroplast if cell just below epidermis. No centrosome.

Nutrition

All living organisms either make or require food. Their types of nutrition can be divided into four basic groups; holozoic, holophytic, saprophytic and parasitic. For course work and in examination questions you may be asked to compare examples of organisms exhibiting different types of nutrition. Useful ones could be selected from the following list.

Holozoic nutrition: Shown by animals which ingest food usually through a mouth, e.g. Amoeba, Hydra, mussel, earthworm, locust, housefly, snail, goldfish, frog, rabbit, cat, cow and man.

Holophytic nutrition: Exhibited by plants which make food by photosynthesis, e.g. Algae (Spirogyra and seaweeds), mosses, ferns, conifers and flowering plants.

Saprophytic nutrition: Shown by organisms which digest organic matter externally and absorb the soluble food through their body surfaces, e.g. many bacteria and fungi (Mucor, yeast, Penicillium and mushroom).

Parasitic nutrition: Exhibited by organisms which live in or on another plant or animal obtaining soluble food from their host, e.g. bacteria causing disease (tuberculosis, pneumonia) plant parasites (dodder and mistletoe), and tapeworm.

Sample question

Describe how the following organisms obtain a supply of food; (a) a named protozoan, (b) Hydra, (c) an earthworm, (d) a named insect. In each case explain how the animal ingests its food. (A detailed account of digestion is not required.)

Framework answer: (a) Amoeba or Paramecium would be suitable examples. Amoeba feeds on smaller, active protozoa, e.g. Colpidium, found in pond water and ditches. This bumps into Amoeba's cell membrane. It appears to stick to this and is then surrounded by a cup-shaped pseudopodial growth. This forms a food vacuole around the prey where digestion takes place.

Paramecium, found in fresh water, feeds on smaller ciliates and bacteria which are swept into the oral groove by the cilia which line it. Food vacuoles form at the base of the oral groove which then move into the cytoplasm. Digestion takes place in the food vacuoles.

(b) Hydra feeds on water-fleas which may be very large compared with the size of Hydra. They are caught by the nematoblasts on the tentacles. These have sticky, clinging or poisoned penetrating threads which can be shot out forcibly when stimulated by taste and water disturbance. They can be used once only and are then replaced. They hold and kill the prey which is moved by the tentacles to the mouth. This opens very wide and the oral cone helps to engulf the food. In the gut, it is digested by enzymes from secretory cells, which are stirred around by flagellate cells. Finally the pieces broken off are engulfed by pseudopodial cells. Digestion is completed in food vacuoles.

(c) The earthworm has a mouth on its ventral surface behind the prostomium. Through this it will ingest large quantities of soil and humus. Lumbricus terrestris, one of the common earthworms, will select leaves which it draws into its burrow. These leaves may either be used to plug its burrow or may be eaten later. Digestion is by enzymes, mainly in the intestine, and is assisted by the grinding action of the soil particles in the gizzard.

(d) Housefly or locust would be suitable examples. The housefly has mouthparts forming a retractable tube called the proboscis. The end of this is flattened to form two hairy lobes. It feeds on almost anything organic, e.g. sugar, milk, sweat, faeces, decaying matter. The proboscis is extended onto the food and enzymes flow down this, spreading out under the lobes. The food is then digested and the soluble matter sucked up through the proboscis.

The locust has a complex system of mouthparts for biting and grinding hard vegetation. The labrum at the front covers the underlying parts. The maxillae feel and taste the food while the mandibles bite and grind the food into small pieces.

Diagrams: These should be included in each section to draw attention to the parts discussed. With (a) either Amoeba shown engulfing food or Paramecium drawn to show the oral groove. In (b) Hydra showing the tentacles and mouth, and an enlarged diagram of a nematoblast. With (c) a diagram to show the mouth, crop and gizzard of earthworm, and for (d) either a diagram of the proboscis of housefly or the mouthparts of locust.

Carbohydrates, Proteins and Fats

The basic food of animals consists of carbohydrates, proteins and fats. These are organic chemicals which are manufactured by plants. By the process of photosynthesis, glucose is produced from the raw materials carbon dioxide and water. In most plants the glucose is converted to, and stored as starch. Glucose and starch are examples of carbohydrates. By a series of chemical reactions simple carbohydrates can be converted into fats. Mineral salts, taken up by the roots of plants, and carbohydrates are required to produce proteins. These organic chemicals are needed to produce the plant tissues.

When the plant tissues are consumed by animals, the plant's proteins, carbohydrates and fats are broken down chemically, by a process known as **digestion**, to soluble substances. The fats are converted to **glycerol** and **fatty acids**, the carbohydrates to **glucose** and the proteins to **amino-acids**. The soluble foods can then be used by the animal for synthesising its own fats, proteins and carbohydrates, to build up its own tissues. These animals may in turn be preyed upon by other animals where the same process will be repeated.

FOOD TESTS

Carbohydrates:

1. Reducing sugars (glucose and fructose), add Benedict's Soln. and boil. A brick-red or orange colour indicates presence.
2. Non-reducing sugars (sucrose), if 1 is negative then add dilute hydrochloric acid, boil for two min and add solid sodium bicarbonate until fizzing stops. Repeat 1. A result as in 1 indicates presence.
3. Starch, add iodine, a blue-black colour gives a positive result.

Proteins:

Boil with Millon's reagent, a pink-red colour indicates the presence of proteins.

Amino-acids:

Boil with Ninhydrin reagent, violet colour indicates the presence of amino-acids.

Fats:

Stir food with a little alcohol. Drop 2 ml. into 2 ml. of water. The presence of fats is shown by a permanent cloudiness.

Sample question – a balanced diet

What are the essential constituents of a balanced diet for Man? Explain the purpose of each substance listed and state which foods would contain them.

Framework answer: Proteins, carbohydrates, fats, vitamins, mineral salts and water are necessary in Man's diet.

Proteins are required for building body tissues and for producing energy. They are obtained from meat, fish, eggs, milk, and cheese. Plant proteins are regarded as inferior since they do not contain all the necessary amino-acids.

Carbohydrates in Man's diet provide the major source of energy and have the advantage of being relatively cheap. Man's main supply is starch from cereals and potatoes, and sugar.

Fats also supply a source of energy but yield nearly twice as much per gram as carbohydrates and proteins. They are found in meat and milk, lard and butter and seeds (maize, groundnut and castor oil).

Vitamins are organic molecules which Man can either not synthesise in sufficient quantities or cannot synthesise at all. They take part in certain essential chemical reactions in the body. Insufficient vitamins can cause various diseases to develop. Vitamin A is required for skin growth and night vision, and occurs in fish-liver oils, milk. It originates from plant carotene. Vitamin B is a group of chemicals involved in energy release in cells. Vitamin B includes thiamine obtained from yeast and wheat germ, riboflavin from yeast, liver, egg, cheese and milk, and niacin from meat, bread and yeast. Vitamin C is needed for skin growth and capillary strength, and occurs in fresh fruit and vegetables. Vitamin D is necessary for incorporation of calcium and phosphorus into the bone, and is found in egg yolk, and fish-liver oils. A similar substance is made by the skin when it is exposed to U.V. rays.

Mineral salts are required to provide certain elements for organic molecules, and to maintain an osmotic balance between cells and fluids in the body. Sodium and potassium are needed for muscles, nerves and plasma and are obtained from salt and plant food, chlorine from salt for muscles, nerves, plasma, saliva and gastric juice, iron from liver and meat for haemoglobin, iodine for thyroxine from plants and water, and calcium and phosphorus from milk for bones and teeth.

About 4 pints of water are required each day. It is needed for protoplasm, body fluids and to help maintain the body at a constant temperature, and to dilute waste to be excreted.

ENZYMES

You will hear and read about enzymes when studying digestion, however it should be remembered that enzymes are of vital importance to any living organism. An enzyme is an organic catalyst. A catalyst is a substance which, in very small quantities, can speed up a chemical reaction without being used up in the reaction. Catalytic processes control the chemical reactions of the cell. These organic catalysts are made by the cells and are always complex proteins. The major difference between an enzyme and an ordinary catalyst is its **specificity**. Each enzyme in the cell is able to catalyse just one specific chemical reaction. This controls the complex chemistry of the cells since the enzymes present will determine the chemical reactions which can take place.

Enzymes are required for respiration, photosynthesis, protein synthesis, digestion and all other biological processes. The enzymes necessary for digestion differ in that they usually function outside the cells where they were manufactured, most other enzymes work in the cells in which they were produced.

Sample question: What are enzymes? State where they are produced in the body of a mammal. Describe experiments to show the properties of any one named enzyme.

Framework answer: Enzymes are biological catalysts, made of protein, and manufactured by cells. They cause chemical reactions to take place rapidly and at much lower temperatures than in their absence, without being used up themselves. They work best at or just above body temperature but are destroyed at higher temperatures.

Enzymes are produced in the cells of the body of a mammal. The enzymes for respiration are produced in every cell. Digestive enzymes are made in the cells of certain glands; salivary, peptic in the stomach wall, intestinal in walls of duodenum and ileum, and in the pancreas.

The enzyme, ptyalin, which breaks down starch is obtained from saliva. Since this is easily collected, it is one of the most suitable for experiments.

Collect saliva, which contains ptyalin, in a test-tube, and dilute with water. Show that saliva is slightly alkaline by testing with neutral litmus paper. Divide the solution into three parts; to the first add some starch solution; boil the second for a few minutes, cool and add starch; to the third add a little dilute hydrochloric acid and then starch. Stand the three test-tubes in a water-bath at 37°C for some time. Add dilute iodine solution to each tube. The three test-tubes are then examined. The first tube shows no colour change; the second and third tubes turn blue/black. Therefore starch breakdown has occurred in the first tube only. Hence, ptyalin digests starch in alkaline solutions, not in acid. Its activity is destroyed by high temperatures. Repetition of the experiment in ice-cold water would show that the enzyme is inactive at low temperatures. Similar experiments could be carried out using other substrates instead of starch, e.g. sucrose, protein. No breakdown of these substances would take place showing that ptyalin is specific for starch.

PHOTOSYNTHESIS

Experiments on photosynthesis

It is useful to bear the following points in mind:

1. Photosynthesis in plants is usually demonstrated by testing for the presence of starch in green leaves. Not all plants store starch temporarily in their leaves so examples for this purpose should be selected from plants with starch-storing leaves, e.g. geranium, Coleus, nasturtium.

2. Use complete plants or leafy shoots in water, not isolated leaves with no water.

3. When illustrating conditions for photosynthesis, (light, a suitable temperature, carbon dioxide and chlorophyll), leaves

must be de-starched first by keeping plants in darkness for 24 – 48 hours.

4. Suitable control experiments should be set up at the same time, e.g. (a) when showing that light is necessary for photosynthesis another plant (control) should be given otherwise identical conditions but placed in the dark, (b) when showing that chlorophyll is necessary for photosynthesis, variegated leaves (i.e. leaves with green and non-green areas) are used in most cases. The non-green areas act as a control. Suitable plants to use would be Coleus, Tradescantia, variegated maple or geranium.

5. In the typical experiment with pondweed (e.g. Elodea) to show the production of oxygen during photosynthesis, if you are making a diagram, take care to avoid some common errors. Do not put the mouth of the funnel on the bottom of the beaker since there should be a free circulation of water. Do not draw the stem of the funnel projecting above the water surface as a tube of water has to be placed over it without losing the water. Remember to show on the diagram that the apparatus is placed in light.

Expt. to show release of oxygen in photosynthesis

General questions on nutrition

1. Name the chief materials required for food manufacture in a green plant. Which of these are absorbed (a) by the roots of plants and (b) by the leaves? What part of the root is mainly concerned with this absorption? What is the importance of this food manufacture to animals?

2. What do you understand by the term photosynthesis? State the conditions necessary for this process and describe one experiment in each case to support two of your statements.

3. Make a large labelled diagram of the alimentary canal of a **named** mammal. Show clearly where (a) pepsin, (b) bile, (c) ptyalin and (d) lipase are produced. Indicate also the places where protein breakdown occurs, absorption of soluble food and water takes place and where the food changes from acid to alkaline.

4. What are carbohydrates? Name three and describe tests to identify them. How does a carnivore obtain a supply of carbohydrates although its food is mainly protein?

5. Describe the dentition of a **named** herbivore and explain how the teeth are adapted to suit its diet.

Framework answers

1. Since the first part of this question refers to food manufacture not photosynthesis, the materials required should include those needed for protein synthesis. Therefore the names will be: **carbon dioxide**, **water** and **mineral salts.**

(a) Water and mineral salts are absorbed by the roots

(b) Carbon dioxide is absorbed by the leaves.

Root hairs are mainly concerned with absorption of water and mineral salts. (A suitable diagram of a root with root hairs should be included here.) Food manufactured by a plant is used to form plant tissues. These are made of proteins, carbohydrates and fats. Animals may obtain their supply of proteins, carbohydrates and fats by feeding directly upon plants. Other animals (carnivores) may obtain theirs indirectly by feeding upon herbivores. Animals cannot manufacture food and therefore depend upon plants for a supply of pre-synthesised food. This food is digested by animals and the soluble food used to build animal tissues.

2. Photosynthesis is the process in which carbon dioxide and water are converted to glucose and oxygen. Other carbohydrates (e.g. starch, sucrose) may be produced from the glucose. The energy required for this is light energy from the sun absorbed by the chlorophyll of plants.

The conditions necessary for photosynthesis are: a supply of carbon dioxide, light, a suitable temperature and presence of chlorophyll.

Two of the following experiments:

(a) To show necessity for carbon dioxide:
De-starch leaves by placing two nasturtium plants in darkness for 48 hours. Absence of starch shown by taking one leaf and immersing it in boiling water to kill the tissues, boiling in alcohol to remove the chlorophyll and then covering leaf with iodine solution. No blue/black colour indicates absence of starch.
Give diagram of suitable apparatus. Place one plant in a bell-jar through which air is passed. Place the other in a similar bell-jar containing a dish of moistened pellets of sodium hydroxide through which air free from carbon dioxide is passed. Expose both to light for six hours. Remove a leaf from each plant and test for starch. (You need to give details of the starch test once in the question.) Blue colour in leaf from first apparatus, i.e. with carbon dioxide, brown in second, i.e. without carbon dioxide. Hence carbon dioxide is required for photosynthesis.

(b) To show necessity for light:
De-starch a nasturtium plant. Cover both sides of part of a leaf with dark paper so that light is completely excluded. Place in light for six hours. Remove leaf and test for starch. The part which was covered remains brown, the portion which was exposed to light turns blue. Hence light is necessary for photosynthesis.

(c) To show necessity for chlorophyll:
De-starch plant of variegated Tradescantia or similar plant. Place in light for six hours. Remove a leaf and map green and non-green areas. Test leaf for starch. Only areas where chlorophyll had been present gives positive test with iodine. Hence chlorophyll is necessary for photosynthesis. Make a diagram of the leaf before and after the starch test.

3. In questions of this type, the answers to each part should be indicated clearly on the diagram. No additional writing is required. Remember to state the name of the mammal drawn.

4. Carbohydrates are compounds of carbon, hydrogen and oxygen. In each molecule there are two atoms of hydrogen for each atom of carbon and oxygen, (CH_2O). Examples are **glucose**, **sucrose** and **starch.**

Tests: To a solution of the substance add an equal volume of Fehling's (or Benedict's) solution. Boil. A red precipitate indicates the presence of glucose.

If no precipitate is obtained, to a fresh solution add a few drops of dilute hydrochloric acid and boil. Cool, then add solid sodium bicarbonate to neutralise the acid. Now repeat the test for glucose.

A red precipitate now indicates that sucrose is present. Boiling the sucrose solution with acid causes the sucrose to be broken down to glucose and fructose. Hence, a positive test is now obtained with Fehling's or Benedict's solution.

Starch is recognised by the fact that it turns dilute iodine solution blue/black.

A carnivore in eating flesh will obtain some glycogen from muscle and liver. Glycogen is a carbohydrate produced from glucose. Normally there is not sufficient of this for the requirements of the carnivore. By digestion of protein, more amino-acids pass into its blood than is needed for synthesis of its own proteins. These excess amino acids pass to cells of the liver and are deaminated, i.e. converted into glucose and ammonia. The ammonia combines with carbon dioxide to form urea which is excreted. Hence, a carnivore obtains carbohydrates by digestion of glycogen to form glucose, and from the excess amino-acids producing glucose in the liver.

Diagram of alimentary canal of human

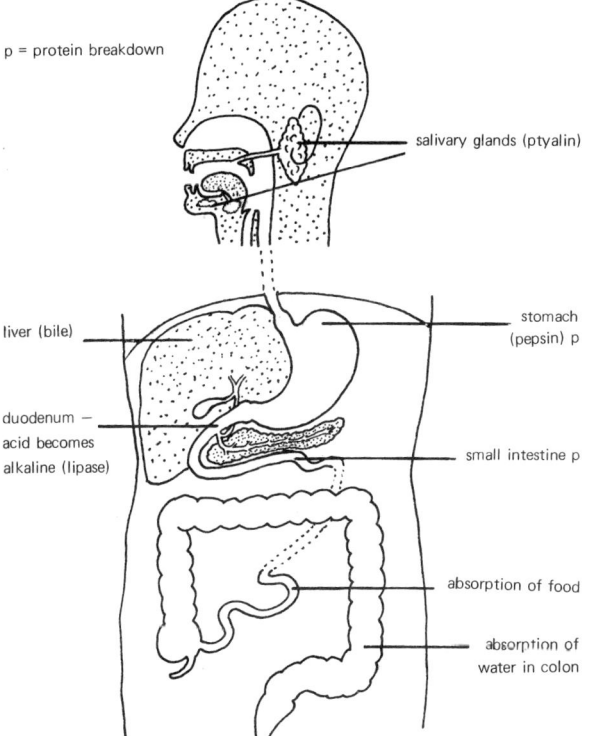

p = protein breakdown

salivary glands (ptyalin)

liver (bile)

stomach (pepsin) p

duodenum — acid becomes alkaline (lipase)

small intestine p

absorption of food

absorption of water in colon

5. The **sheep** or **rabbit** would be suitable examples of herbivores.
Sheep: The dental formula for sheep is:

$$i. \frac{0}{3}; \ c. \frac{0}{1}; \ p.m. \frac{3}{3}; \ m. \frac{3}{3}$$

This shows numbers of different types of teeth in upper over lower jaw for each half jaw. The sheep has six incisors (i.) in the lower but no incisors in the upper jaw. The incisors are chisel-shaped teeth used for biting off pieces of food. Canines (c.) are pointed teeth for tearing or stabbing flesh. In sheep these are not required for this purpose. The incisors and two canines in the lower jaw are all alike but are missing in upper jaw where a horny pad is used as a chopping board to grip and break off pieces of grass. The premolars (p.m.) and molars (m.) are large teeth with flat tops bearing small points or cusps. They are used for crushing and grinding food. The sheep has six molars and six premolars on the upper and lower jaw. These move sideways against each other to grind grass into small pieces. This is necessary since the essential food in the grass is trapped inside the plant's cellulose cell walls. This is released by the grinding action of the teeth on the grass to break the cell walls. The teeth of sheep are subjected to considerable wear and to counteract this they grow continuously.

Rabbit: The dental formula for rabbit is:

$$i. \frac{2}{1}; \ c. \frac{0}{0}; \ p.m. \frac{3}{2}; \ m. \frac{3}{3}$$

There are four upper incisors arranged in two pairs, one pair behind the other. The front pair are long, curved teeth with thick enamel at the front. This wears less than the other parts of the teeth so producing a sharp edge. The single pair of incisors in the lower jaw have a similar sharp edge, and the upper ones work against them. The canines are absent in rabbit and there is a gap known as the diastemma. Behind this there are six molars and six premolars in the upper jaw and four premolars and six molars in the lower jaw.
The incisors bite off pieces of grass which are then passed back to the cheek teeth, where they are ground by the backwards and forwards movement of the lower teeth on the teeth of the upper jaw. The upper jaw is hinged very loosely so allowing this movement.

28

Respiration

Every living cell carries out respiration. This usually involves the breakdown of carbon compounds, particularly glucose or similar molecules, to release **energy**. This takes place in the mitochondria of the cells. Respiration should not be confused with **breathing** which refers to the method by which air is drawn in and out of the lungs.

Importance of respiration: This frequently forms part of an examination question. Respiration is important because the energy produced is essential to a living organism. The energy released by the breakdown of organic matter is partly lost in the form of heat and the remainder is stored chemically by the formation of molecules of adenosine triphosphate (ATP). ATP can be used by the organism when energy is required. Energy is needed for:

(a) Synthesis of organic molecules, e.g. enzymes (proteins) and proteins for building new tissues. This is true for all living organisms, plant and animal.

(b) Movement, voluntary movements, e.g. walking, swimming and involuntary movements, e.g. heart beat, breathing.

(c) To maintain the correct environment in the cells.

Internal and external respiration: The respiration which takes place in all living cells is referred to as **internal** respiration. When glucose is broken down oxygen is usually required and carbon dioxide and water are usually released. The method by which organisms obtain their oxygen (if required) and release their carbon dioxide and water is referred to as **external** respiration. Thus breathing is part of external respiration.

INTERNAL RESPIRATION

Aerobic respiration. This process occurs in the presence of oxygen. Glucose is broken down to release carbon dioxide and hydrogen which combines with oxygen to form water. (This form of respiration produces a large amount of energy, 650 Kcal per gram mol.) It can be summarised by the following equation:

$$C_6H_{12}O_6 + 6O_2 \rightarrow 6CO_2 + 6H_2O + energy$$

29

This is only an overall picture of what occurs. The process is highly complex taking place in many stages and not as a single step as the equation indicates. Aerobic respiration produces a large amount of energy.

Anaerobic respiration. In the absence of oxygen, anaerobic respiration takes place. A few types of organisms use this form of respiration as their normal method, e.g. yeast, certain bacteria. It can, however, occur in plant and animal tissues when the oxygen level is low. Instead of the complete breakdown of carbon compounds, which takes place in the presence of oxygen, only partial breakdown occurs with the accumulation of certain by-products.

Examples. (a) **Yeast,** a type of fungus, obtains its energy by the breakdown of glucose to carbon dioxide and alcohol.

$$C_6H_{12}O_6 \rightarrow 2CO_2 + 2C_2H_5OH + energy$$

(b) **Bacteria in milk** break down the sugars in milk to lactic acid. This is the cause of milk turning sour.

$$C_6H_{12}O_6 \rightarrow 2C_3H_6O_3 + energy$$

(c) Alcohol may accumulate in plant tissues when the oxygen level is low, and lactic acid in animal tissues under the same conditions.

Experiments

Ideally, to show that a living organism respires, one should show experimentally that chemical energy is released from the glucose molecules. However, this is not possible by simple experiments, hence gaseous exchange is observed instead. It is easier to show the release of carbon dioxide than the uptake of oxygen, therefore when demonstrating that living organisms respire, it is usual to observe or measure the production of carbon dioxide. This is detected by the reaction with **limewater** or **bicarbonate/indicator solution**. Carbon dioxide turns limewater chalky. The bicarbonate/indicator solution contains a dye which is red when air or oxygen is bubbled through but turns yellow if carbon dioxide is passed through. This solution can be used for comparing rates of respiration by measuring the time taken for the dye to change colour.

1. To show that carbon dioxide is present in exhaled air:
The apparatus consists of two wash-bottles joined by a T-junction
at which you breathe air in and out. Either limewater or bicar-
bonate/indicator solution can be used.

Result: The solution through which atmospheric air passed
remained unchanged while the exhaled air caused either the
limewater to turn chalky or the bicarbonate/indicator solution to
turn yellow. This shows the presence of carbon dioxide in exhaled
air.

Expt. to show that carbon dioxide is present in exhaled air

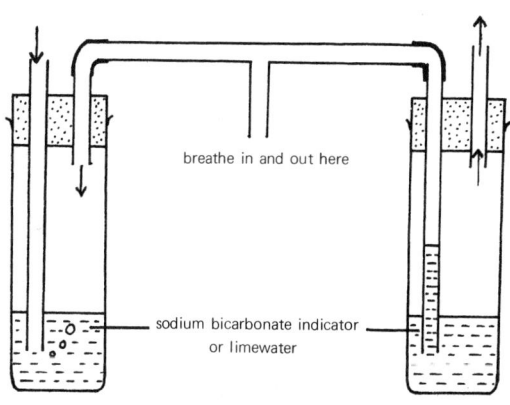

breathe in and out here

sodium bicarbonate indicator
or limewater

2. To show that living organisms produce carbon dioxide:
Set up the apparatus as shown with 2cm³ of bicarbonate/indicator in the tube. Small animals, e.g. locust hopper, woodlouse or non-green parts of plants, e.g. germinating seeds, can be placed on the platform.

Result: After a period of time the solution turned yellow, this showed that the organism in the tube had produced carbon dioxide.

3. To show that respiring peas give out heat (see diagram opposite). The peas in flask A are soaked in formalin solution. This is poison which will kill the peas but prevent bacterial decay. The temperatures are noted each day for a week.

Result: The temperature in flask A remained unchanged while the temperature in flask B went up by several degrees, thus demonstrating that respiring peas give out heat.

4. To demonstrate anaerobic respiration:
(a) In germinating peas: Fill a test-tube with mercury and invert this into a small dish also filled with mercury. The testas are removed from a number of germinating peas to ensure that air is not trapped inside. The peas are then placed under the opening to the test-tube so that they float to the top. This procedure ensures that the peas are without oxygen.

Result: After a few days a gas had collected around the peas. When removed from the mercury and tested with limewater it was found to be carbon dioxide, thus showing that respiration had occurred in the absence of oxygen.

Experiment to show that living organisms produce carbon dioxide

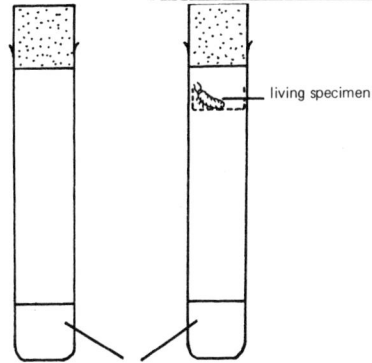

living specimen

sodium bicarbonate indicator

Experiment to show that respiring peas give out heat

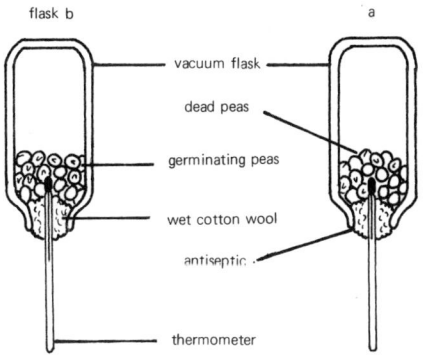

flask b

a

vacuum flask

dead peas

germinating peas

wet cotton wool

antiseptic

thermometer

(b) In yeast: Take two test-tubes and into each put 5 cm³ of glucose solution. To the first add 1 cm³ of yeast suspension and to the second 1 cm³ of boiled yeast suspension. Then fit balloons over each test-tube and place in a water-bath at 37°C observing the results after 60 minutes.

Expt. to show anaerobic respiration in yeast

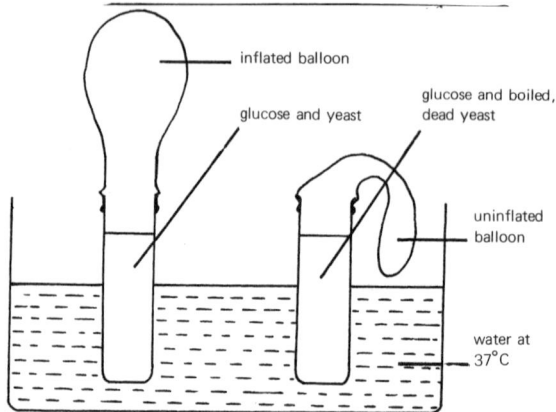

Result: the balloon over the first tube had inflated, and when tested with limewater, was found to contain carbon dioxide. The balloon over the second tube remained deflated which showed that the boiling had inactivated the yeast. The liquid in the first tube smelt of alcohol showing that fermentation had taken place.

Note

In these respiration experiments, if you use peas as your respiring material, the boiled (dead) ones must be soaked in formalin which kills the bacteria which would bring about decay. These bacteria respire and would make the control part of your experiment inaccurate.

General Questions on Respiration

1. What is the importance of respiration? Describe how oxygen gets to the tissues of a mammal.

2. Explain why all living organisms respire. How do three of the following obtain a supply of oxygen; (a) Amoeba, (b) an adult insect, (c) a fish, (d) an adult frog.

3. Distinguish between breathing and respiration. What is the difference between inhaled and exhaled air? Describe briefly simple experiments to demonstrate these differences.

4. Equal quantities of peas were placed into three vacuum flasks. These flasks were inverted and their necks plugged with cotton wool through which passed thermometers: Flask A contained germinating peas, flask B boiled peas and flask C contained boiled peas soaked in formalin. Temperature readings were taken every day for five days as shown below.

DAY	FLASK A	FLASK B	FLASK C
0	20°C	20°C	20°C
1	20·5°C	21°C	20°C
2	21°C	22·5°C	20°C
3	22·5°C	24°C	20°C
4	23°C	25·5°C	20°C
5	24°C	27°C	20°C

Answer the following questions:
(a) What vital process does this experiment show?
(b) Why has the temperature increased in flask A?
(c) Why has the temperature increased in flask B?
(d) What is the purpose of flask C?
(e) What do you think will happen to the composition of the air during the five days?

Answers

1. All living organisms must have energy available to carry out chemical changes and movement. The energy is released when organisms respire. (See page 29).
The answer to the second part of the question can be divided into three parts; (a) the passage of air into the lungs, (b) the diffusion of oxygen into the blood stream and (c) the transfer of oxygen to the tissues.

(a) A diagram of the thorax with arrows to indicate the route taken by air, should be given.
When air is inhaled, the muscles between the ribs contract pulling the ribs outwards and upwards, and the dome-shaped diaphragm contracts and flattens. This enlarges the volume of the thorax which in turn decreases the internal pressure. Thus air passes into the lungs.

(b) A diagram should be included to show the blood capillaries around the air-sac.
The oxygen in the air-sacs or alveoli diffuses into the mucus which lines the walls. From here the oxygen passes across the walls of the alveoli and into the blood capillaries which surround the alveoli. This diffusion of oxygen takes place because there is a higher concentration of the dissolved gas in the air-sacs than there is in the blood capillaries. Since the membranes between them are permeable to oxygen it will diffuse across.
In the blood, the oxygen combines with haemoglobin in the red blood cells to form a bright red pigment oxy-haemoglobin. The oxygen is transported in this form to all parts of the body, passing through the heart first so that the blood can be pumped to all tissues.

(c) The blood passes in large blood vessels, called arteries, to the tissues. Here the blood vessels become much smaller with thin permeable walls. These blood vessels are known as capillaries. During the passage of blood through these blood vessels, oxygen diffuses into the tissues.

2. See the answer to the first part of the previous question. Three of the following should be selected.

(a) Amoeba. This organism has a large surface area compared with its volume. Hence Amoeba can obtain sufficient oxygen by simple diffusion. Oxygen dissolved in the water in which it lives diffuses into the Amoeba cell over its entire surface.

(b) Adult insects have a series of openings, spiracles, along the sides of their thorax and abdomen leading to a system of air tubes, trachae, which end in the various body tissues. Movements of the body, especially the abdomen, cause a pumping action so that air flows in and out of the tubes. Oxygen then dissolves in the tissue fluid.

(c) Fish obtain a supply of oxygen through their gills. These are situated in the gill cavity. The gills consist of thin finger-like processes enclosing a supply of blood capillaries. Water, containing dissolved oxygen, enters the fish through the mouth when the floor of the mouth is lowered. The water passes back over the gills where oxygen passes by diffusion into the blood. Oxygen is then carried to all tissues of the fish in its blood vessels.

(d) Adult frogs use their skin, mouth cavity lining and lungs as respiratory surfaces. These are all moist and well supplied with blood vessels. Some of the oxygen is always obtained by diffusion through the skin. When the frog is sitting still, the rest of the gas is absorbed in the mouth. Air is drawn in by lowering the floor of the mouth causing air to enter through the nostrils, the mouth being closed. This air is forced out through the nostrils by raising the floor of the mouth. When in need of extra oxygen, air is forced into the lungs by closing the nostrils and raising the floor of the mouth.

4. (a) That living organisms generate heat.

(b) Peas undergoing respiration give off heat.

(c) Decay organisms feeding on dead peas respire giving off heat.

(d) Control. Peas are dead, therefore not respiring. They are sterile therefore there are no bacteria respiring in the flask.

(e) A, B increase in carbon dioxide concentration, decrease in oxygen. C unchanged.

Transport

Materials are transported around the cell, also into and out of the cell through the living cell membranes which take part in the process and around the bodies of multicellular organisms. It is important to remember that substances may be moved even though the organism may not use any energy, for example the diffusion of carbon dioxide to the exterior from Amoeba, or transport may be active as it is when the same animal gets rid of water against the concentration gradient.

Organisms need to transport food, water and mineral salts for energy and growth, and oxygen for respiration. To maintain homeostasis, excretory products must be removed from the cell. In multicellular organisms, hormones (chemical messengers) need to pass from where they are made to where they cause an effect. The transport of water or blood helps to keep body temperature constant, in addition, blood carries defensive chemicals and cells.

TRANSPORT IN FLOWERING PLANTS

Water: This forms an essential part of protoplasm; maintains turgor; acts as a solvent for transport of materials; provides hydrogen atoms for synthesis of carbohydrates in the process of photosynthesis.

Water is absorbed by the root hairs by osmosis. It is transported in the xylem, also by osmosis. As well as osmosis, root pressure and transpiration also cause movement of water in the plant.

Transpiration is a process whereby plants lose water as water vapour into the atmosphere. It takes place mainly through the leaves. Water from the cells passes through the cell walls into the intercellular spaces. The water diffuses out into the atmosphere through pores, the stomata. These are mainly found in the epidermis of the lower surface. Using an atmometer it can be shown that water evaporates at different rates in different atmospheric conditions. These external conditions also have an effect on the evaporation of water from the plant, i.e. transpiration rate.

The atmospheric conditions which have an effect on transpiration are, degree of saturation of the air, temperature and air currents. The continual loss of water from the leaves causes water to be sucked up through the stem from roots and finally from the soil. This stream of water is the transpiration stream.

Atmometer

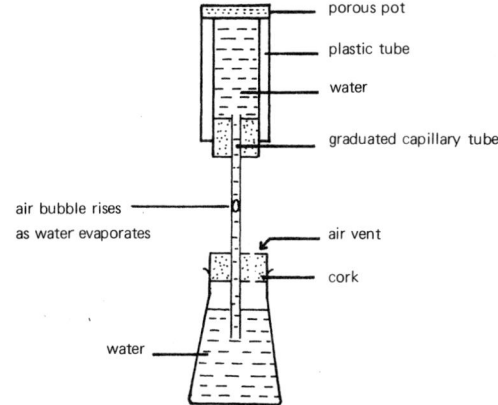

porous pot

plastic tube

water

graduated capillary tube

air bubble rises
as water evaporates

air vent

cork

water

Experiments

1. To show the use of xylem for the transport of water: Cut a stem of white deadnettle under a solution of red dye (eosin). Leave. The veins in the flowers will go red and the stem, if cut and examined with a hand lens, will show dye in the xylem only.

2. To show uptake of water by a leafy shoot: The potometer is completely filled with water and set up as shown. The meniscus moves as water is drawn from the apparatus to replace that lost by evaporation from the leaves. The rate of water uptake as a result of transpiration, can be found by timing the movement of the meniscus over a fixed distance.

The stem must be cut and put into the tube under water to prevent air blocking the xylem. It is essential that the whole apparatus is air tight. It can be used to show the effects of different temperatures and air currents on transpiration. The shoots in the apparatus must be left for some minutes in the new conditions to adapt before readings are taken.

Potometer

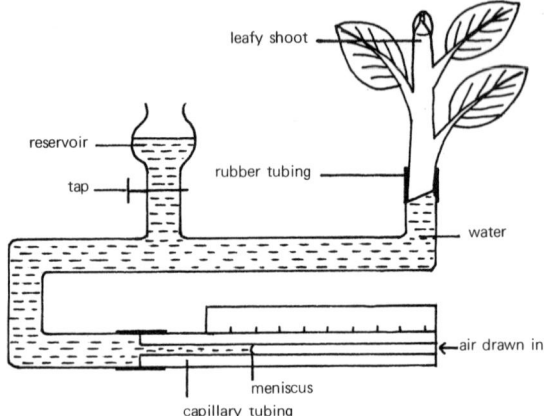

3. To show the effects of light on stomatal aperture: Place a drop of benzene on a leaf which is brightly illuminated. The benzene enters the leaf making it look dark in this area. Another drop of benzene placed elsewhere on the same leaf, but in darkness, does not enter. This indicates that the stomatal aperture becomes reduced in darkness.

Stomatal aperture (stereogram)

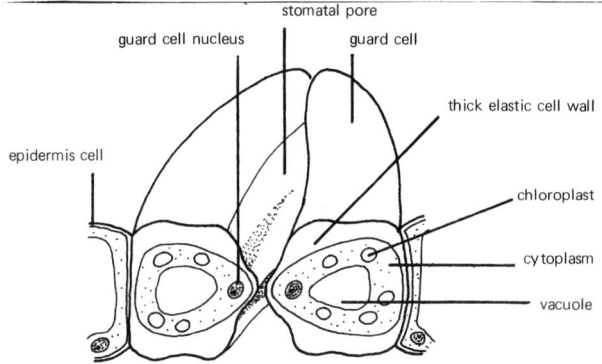

Root pressure experiment

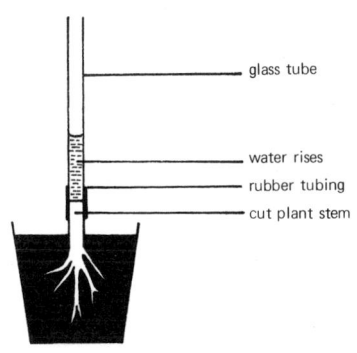

Bark ringing experiment

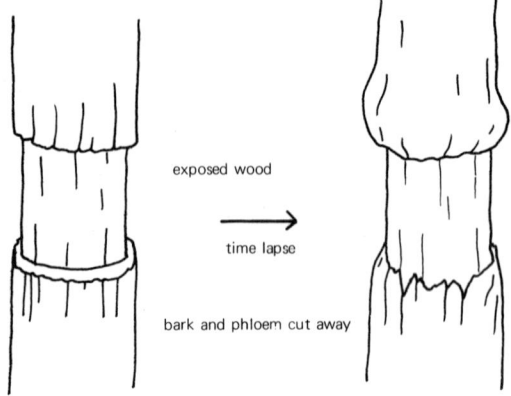

exposed wood

time lapse

bark and phloem cut away

Salt transport: Salts do not enter by osmosis which is the movement of water only. They are not absorbed by diffusion alone as soil water is often more concentrated than the plant cell. Salt uptake stops if the cells are deprived of oxygen, thus it must be an energy requiring process. Salts reach xylem in a similar way, and then travel with water.

Food transport: This takes place in the phloem, and is an energy requiring process and can only take place in living cells. Aphids feed by sticking a tube (proboscis) into the stem of the plant and withdraw food from the phloem. If they are killed, and their bodies removed, leaving the proboscis still in the stem, fluids coming from the phloem are found to be rich in sugars and amino acids. The diagrams above show the effects of bark ringing. This removes the phloem thus preventing the transport of food to the roots of the tree. The swelling above the ring is due to extra growth caused by the accumulation of food. The tree eventually dies as the roots are deprived of food.

General question

Transpiration experiments do not actually demonstrate the process, but usually investigate it by measuring uptake or loss of water. Whole plants, shoots or leaves may be used. For this reason you must read the questions on this topic very carefully. Candidates often lose marks by describing the incorrect experiment.

1. Below is a diagram of an experiment using a weight potometer. Answer the following questions about it.

(a) What is the function of the oil?

(b) When water is added to bring the level in the flask up to the mark, the plant is found to be heavier than it was at the beginning of the experiment. How do you account for this increase in weight?

(c) The result in this experiment would indicate that the plant is losing water to the atmosphere. Give one reason why this experiment cannot give an absolutely accurate result of the amount of water lost by transpiration.

(d) How could you use this apparatus to find out if leaves transpire all over their surface? You may use several flasks.

Diagram of weight potometer

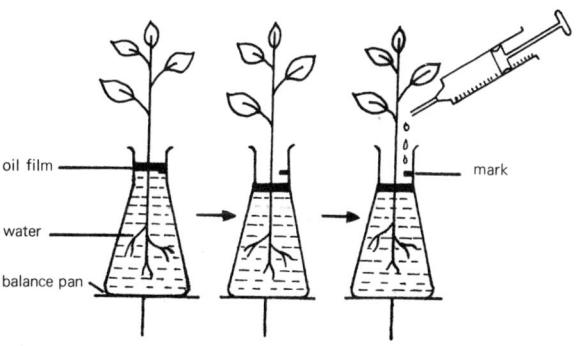

add water of known volume

oil film

water

balance pan

mark

2. Describe an experiment you have done using living plant material to demonstrate the process of osmosis. By means of a large labelled diagram show how water enters the root hair by osmosis.

3. Describe one experiment in each case to show (a) that water vapour is lost from leaves, (b) the rate of water loss from a cut shoot.

4. Describe an experiment for measuring the rate of water absorption by a shoot. What part does water play in the life of the plant? What changes in external conditions would cause an increase in absorption of water?

Framework answers

1. (a) Prevents evaporation of water directly from flask, hence reduces experimental inaccuracy.

(b) Photosynthesis has been taking place so plant has been making food, hence increase in weight.

(c) Some water used in photosynthesis, so not all water absorbed is given off in transpiration.

(d) Set up 4 cut shoots immersed in water in graduated cylinders with layer of oil on top of each. Use plants with simple leaves (e.g. Willowbay herb). Brush onto the leaves of the shoots a coating of vaseline/paraffin mixture as follows:
> A upper leaf surfaces
> B lower leaf surfaces
> C both leaf surfaces
> D leave untreated as control

Result: Water loss in order of magnitude, D, A, B, least C. Therefore the leaves lose most water from the lower surface.

2. Cut three half potatoes, making a cavity in each. Boil one to kill protoplasm. Place them in petri dishes with water. Put salt into the cavity of the boiled and one raw potato. After two hours water has collected in the cavity of the raw potato indicating osmosis has occurred. The dead potato is the control. (A diagram of the experiment should be included here.)

3. (a) Using a potted plant, enclose pot and soil in plastic bag. After watering soil, cover with bell-jar. For control use a similar plant from which leaves have been removed. Leave. Inside walls of first bell-jar become covered in droplets of colourless liquid, the other does not. Liquid identified as water as it turns dry blue

44

cobalt chloride paper pink. Therefore water vapour, which must have come from the leaves, had condensed on the bell-jar walls.

(b) Place shoot in tube of water, covering surface with a thin film of oil, suspend from balance, weigh at intervals. Then calculate loss in weight per hour.

4. Potometer set up using a freshly cut shoot. This filled with water and made airtight. The movement of the meniscus over a set distance, as the shoot absorbs water, shows the rate of water absorption. A diagram of the apparatus, fully labelled, should be drawn here. Water forms much of protoplasm of cells; maintains turgor; forms solvent for transport of materials; provides hydrogen necessary to form carbohydrates in photosynthesis.

An increase in transpiration causes an increase in water absorption, therefore conditions affecting transpiration will have the same effect on absorption. In the light, dry, warm air which is moving, causes an increase in absorption.

Osmosis experiment

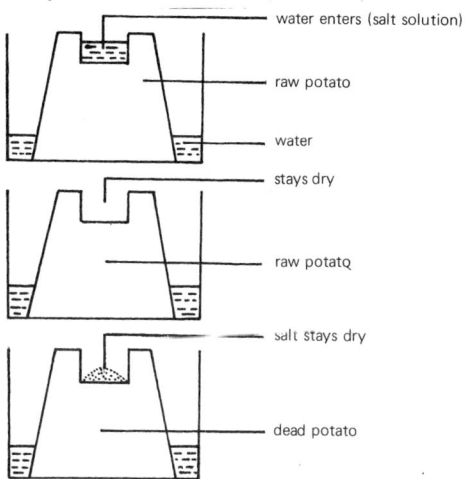

TRANSPORT IN ANIMALS

Blood is the transport system in animals. It consists mainly of plasma, which is a yellowish fluid, and is principally water, contains salts, dissolved food, etc., and proteins. Carried in the plasma are red blood cells containing respiratory pigment haemoglobin. The other blood cells are the white cells concerned with protecting the body. Also in the plasma are platelets; these are tiny protoplasmic particles concerned with clotting. Opposite is a diagram of the double circulatory system in mammals. Note that the blood is pumped to the organs in arteries, and returns directly to the heart in veins. The exception is the hepatic portal vein, which transports blood from the stomach and intestines to the liver. From here it passes in the hepatic vein to the vena cava, then to the heart. Arteries normally carry oxygenated blood, the veins deoxygenated; this is reversed in the pulmonary arteries and veins. The renal artery is rich in waste products, which are removed by the kidney so that blood in the renal vein has a low concentration of wastes. The latter blood vessels are different from other arteries and veins as arteries usually have a much lower concentration of waste products than veins.

If a question asks for a diagram showing only part of the circulatory system, use only the relevant part of the opposite diagram.

General questions

1. (a) Describe the structure of blood and lymph. (b) Write an account of the way in which the heart maintains the circulation of blood. Illustrate your account with simple labelled diagrams.

2. (a) Make a large labelled diagram to show the structure of a mammalian heart and the origins of its main vessels. Indicate by means of arrows the direction of flow of oxygenated blood.

(b) Describe blood cells and state how they carry out their functions.

3. (a) Trace the path of the red corpuscle as it passes from the small intestine to the lungs. (b) Name the substances which are transported by the blood, stating where they are produced and where they are taken.

46

Circulation

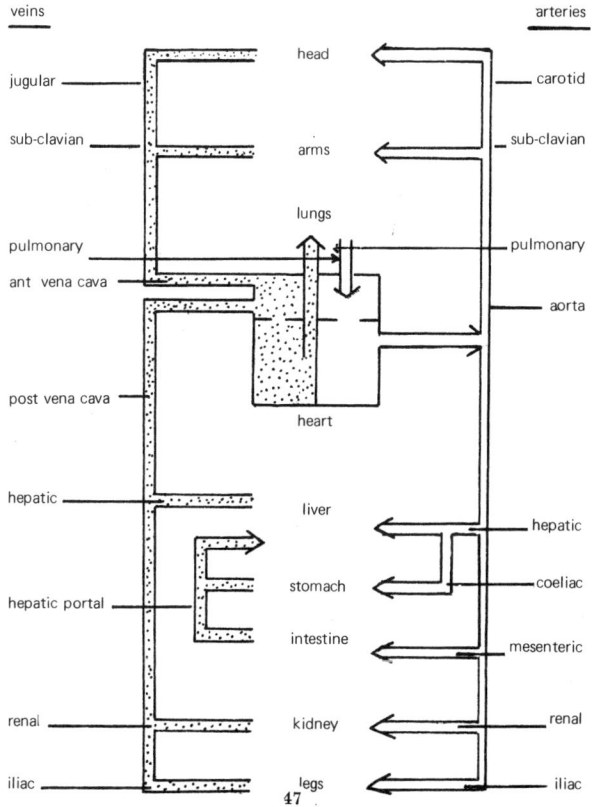

veins · arteries

jugular — head — carotid

sub-clavian — arms — sub-clavian

lungs

pulmonary — — pulmonary

ant vena cava — — aorta

post vena cava — heart

hepatic — liver — hepatic

— — coeliac

hepatic portal — stomach

intestine — mesenteric

renal — kidney — renal

iliac — legs — iliac

Circulation in mammal

4. List the differences between arteries and veins. How is the structure of capillaries related to their function?

Framework answers

Avoid the use of the terms pure and impure blood. Questions are not always as straightforward as the ones given here, but these cover the range of types. At first glance, a question may appear to ask for detail you have never been given, but if you think carefully you will find that you do have the information, it is being asked for in a new way.

1. (a) Blood, a fluid tissue, consists of red blood cells, a number of different types of white blood cells, platelets, all are in liquid plasma. Red b.cs.; minute, biconcave discs of cytoplasm, no nucleus, elastic cell membrane, 5 million/mm³. They are short lived, up to 3–4 months. Contain haemoglobin. White b.cs.; 1 to 500 red b.cs., several types, majority amoeboid, all have a nucleus. Platelets; minute, irregularly shaped pieces of cytoplasm, derived from nucleated cells. Plasma forms just over half the blood volume, pale yellow in colour, 90% water, rest dissolved glucose, amino-acids, salts, vitamins, hormones urea etc. Carbon dioxide and plasma proteins in colloidal state, e.g. fibrinogen, antibodies. Lymph, similar to plasma but less protein. Lymph flows in lymph vessels, it contains white b.cs., lymphocytes, these produce antitoxins and are made in lymph glands.

(b) Heart is four chambered, muscular double pump. When the auricles are full, contraction of both starts simultaneously at anterior end around entrance of veins. It passes down towards ventricles, tricuspid and bicuspid (mitral) valves open, ventricle muscles relax, blood enters ventricles. Contraction of ventricle muscles starts at apex (posterior end), it passes anteriorly, the pressure of blood closes tricuspid and bicuspid valves. Semilunar valves in entrance of pulmonary artery and aorta open. Blood pumped out to arterial system. As left auricle and ventricle are completely separate from the right ones, no mixing of oxygenated and deoxygenated blood occurs. Heart beat rate can be varied, slowed down by nervous stimulation via vagus nerve from brain. Accelerated by stimulation via sympathetic nerves and by an increase in amount of adrenalin in blood.

Diagrams of heart beat

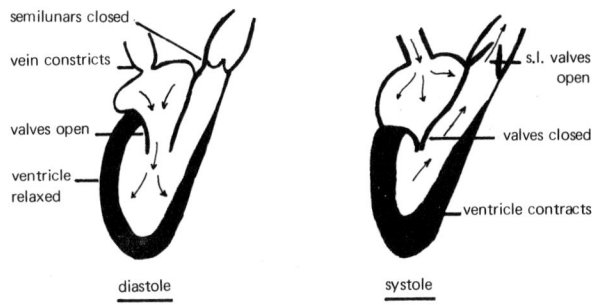

semilunars closed
vein constricts
valves open
ventricle relaxed

diastole

s.l. valves open
valves closed
ventricle contracts

systole

2. (a) Draw a large diagram as simple as possible without missing out essential detail.

(b) See 1 (a)
Functions: Red b.cs.; carriage of oxygen. Haemoglobin forms unstable compound, oxy-haemoglobin, which combines with oxygen as blood circulates through lung capillaries. Oxygen released to cells with low oxygen content. Some carbon dioxide also carried combined with haemoglobin, but most carried in solution in plasma.

White b.cs.; defence of body against bacteria and their toxins. Amoeboid ones ingest bacteria in same way as Amoeba feeds, also help healing by removing dead tissue. Others produce antitoxins which neutralise poisons produced by bacteria, and antibodies which inhibit the activity of bacteria. Some white b.cs. produce enzymes to weaken bacteria before ingestion. Diagrams of red and white b.cs. should be included here.

Platelets; concerned in blood clotting. Exposure to air or damaged tissue causes platelets to break up, releasing an enzyme, this causes release of enzyme thrombin, from prothrombin present in plasma, in the presence of calcium in plasma. Thrombin converts soluble fibrinogen to insoluble fibrin. This forms meshwork trapping red b.cs. to make clot.

3. (a) Intestine, hepatic portal vein, capillary network of liver, hepatic vein, inferior vena cava, right auricle (or atrium), right ventricle, pulmonary artery, capillary network of lungs.

(b) Oxygen from air in lungs to tissues. Carbon dioxide from tissues to lungs. Soluble food from intestines to tissues. Waste products from tissues to kidney. Hormones from ductless (endocrine) glands carried all round body, have an effect on certain organs only. Heat from physically and chemically active organs (e.g. muscles, liver) all round the body.

4.

Arteries	Veins
Thick muscular wall	Thin limp wall
Narrow bore	Wide bore
Valves at start only	Valves throughout
Lie deep in body	Lie near surface
Carry blood at high pressure	Carry blood at low pressure
Blood usually oxygenated	Blood usually deoxygenated
Blood usually rich in food	Blood usually rich in wastes

Function of capillaries is to bring blood as near as possible to cells of the body, and allow diffusion of soluble substances from blood to tissue fluid bathing cells, and to permit waste products to pass in the opposite direction. This is possible because their walls are extremely thin, composed of flattened cells in a single layer.

Excretion

Waste products are those substances made by an organism, undigestable material is not a waste product. Waste substances must be expelled, since if they accumulate they upset the balance of chemicals in the cells (homeostasis), and they may also be toxic.

In simple organisms like Amoeba, Spirogyra, Hydra, etc., waste products diffuse out of the cells into the surroundings. In mammals, waste products are expelled from a number of organs, but the principal organ of excretion is the kidney. This organ filters blood removing urea and excess salts and water. The composition of the material expelled from the kidneys varies with food and water intake, and the body's activity. Thus the kidney regulates the composition of the body fluids and therefore homeostasis.

General questions

1. Draw a labelled diagram of the urinary system of a named mammal. Describe a mammalian kidney and the way in which it functions.

2. Name the excretory organs of a mammal. Outline the ways in which they function, indicating the changes undergone by blood passing through them.

Framework answers

1. Diagram overleaf. In order to describe how the kidney functions it is necessary to draw a kidney tubule, as shown on the next page. Remember that the kidney uses a considerable amount of energy, this is because the reabsorption of glucose and other useful substances in the 1st convoluted tubule is an energy requiring process. So also is the secretion of water and salts into the final part of the tubule from the blood. Kidney is slightly flattened red-brown body, convex on one side, concave on the other. Ureter, renal artery and vein connected to middle of concavity. Consists of vast number of tubules surrounded by capillaries connecting up branches of renal artery and vein. Outer zone, the cortex is made up of all of kidney tubules, except loops of Henlé. It appears darker and denser than inner medulla; in this are tubule loops and collecting ducts leading

to pyramids which open into pelvis. This is a cavity leading to the ureter. There is a network of capillaries, glomerulus, surrounded by the Bowman's capsule at the closed end of each tubule.

Kidney tubule

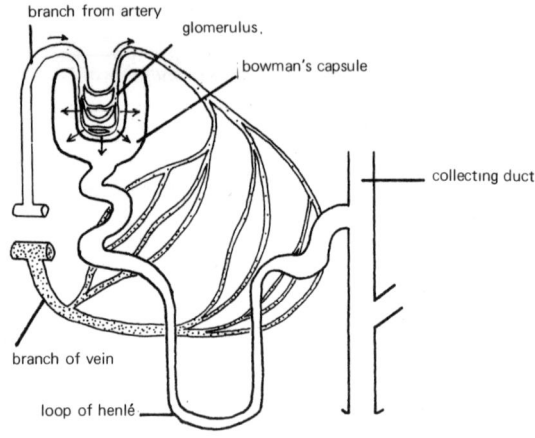

Urinary system of man

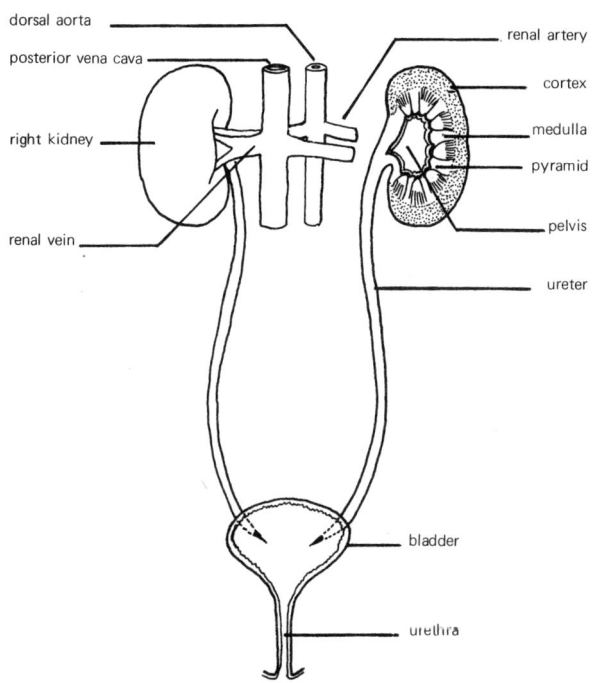

dorsal aorta

posterior vena cava

right kidney

renal vein

renal artery

cortex

medulla

pyramid

pelvis

ureter

bladder

urethra

Functioning: Blood under pressure loses some plasma as passes through glomerulus, proteins do not pass out. Hence there is steady flow of liquid along tubule. In 1st part of tubule, useful substances, e.g. glucose, actively reabsorbed back into blood. Loop of Henlé, water reabsorbed by blood (by osmosis). Amount absorbed depends on water content of blood, which the kidney therefore keeps constant. Final part of tubule, toxic substances and some salts pass from blood into tubule, resulting liquid is urine. This completely different in composition from fluid filtered into Bowman's capsule. Urine passes to collecting duct, pelvis and ureter to bladder for temporary storage.

2. Excretory organs: Kidneys, lungs, liver, skin and to very minor extent, rectum.
Kidney, see question 1 for functioning.

Blood loses food and oxygen in respiration of kidney, needs energy for active reabsorption of some substances. Also loses water, urea and other waste nitrogenous substances, some salts and toxic materials. Gains carbon dioxide.

Lungs. Blood entering lungs contains carbon dioxide in solution in plasma, also bicarbonate ions. Carbon dioxide diffuses through capillary and air sac walls into air sacs. It is released from bicarbonate ions by enzyme action. Efficiency of the lungs depends on their enormous surface area.

Blood loses carbon dioxide and some water. Gains oxygen.

Liver. Nitrogenous part of excess amino acids removed, converted to urea, removed by blood. Bile formed from salts and pigments from decomposition of haemoglobin, passes into the alimentary canal.

Blood entering via hepatic artery, loses oxygen. Blood entering in hepatic portal vein, loses glucose, amino acids. Gains urea, carbon dioxide, if blood sugar low, may also gain glucose.

Skin. Sweat produced by sweat glands, contain traces of urea, much more if kidney diseased.

Rectum. Only an excretory organ in that bile pigments and dead cells from gut lining passed out with faeces, which are not waste products.

Blood in walls gains carbon dioxide and water, loses oxygen.

Temperature Control

All chemical work of the cell is carried out by enzymes which are affected by temperature, therefore this is important to living organisms. If it is extremely high or low, it causes death of most organisms. Different plants and animals are adapted to live at different temperatures.

Plants and cold blooded (poikilothermic) animals have roughly the same temperature as their surroundings. A plant can do little to regulate its temperature. If it is hot, transpiration increases. Water evaporating from the leaves does so using latent heat from the leaves, which has a cooling effect on the plant. Cold blooded animals avoid extremes of temperature by seeking out places with a favourable temperature. Reptiles such as lizards bask in the early morning sun absorbing more radiation than their surroundings. Later in the day when it is very hot, they shelter in shade, so are cooler than their surroundings. Butterflies, if cold, vibrate their wings, the muscular activity releasing heat so preventing body temperature becoming too low. Long periods of cold weather make life impossible for insects. They overcome winter by spending it in a resistant form as a chrysalis, or in a dormant state.

Warm blooded (homothermic) animals maintain a constant body temperature regardless of the temperature of the surroundings. Reducing heat loss: Mammals and birds are both warm blooded. If they are hotter than their surroundings they will lose heat by conduction, convection and radiation. As an insulating layer, many mammals (not man), have fur or hair, birds have feathers (also for flight). Special muscles pull each hair/feather upright. This has the effect of trapping a thicker layer of air next to the body. Air is a bad conductor, therefore reduces heat loss. When frightened or angry animals raise their fur to make them look bigger. Fat beneath the skin is also a good insulator, it is very thick in some aquatic mammals, such as the now almost extinct whale.

Sweating, if it occurs, as it does in man, is reduced to a minimum. In cold conditions, capillaries in the skin contract (vasoconstriction), reducing the amount of blood passing near the surface. Blood also returns to the heart mainly in veins lying deeper in the tissues. Hence the amount of heat lost from the blood by radiation is reduced. If cold, the body also generates more heat by increasing its metabolic rate, and by shivering.

Increasing heat loss. Skin capillaries dilate (vasodilation), much blood returns to the heart in surface veins. Thus much blood passes near the air losing heat. Sweating in man is increased. The sweat evaporates using latent heat from the body, so has a cooling effect.

Question

Make a labelled diagram to show the structure of mammalian skin. State briefly the functions of the parts you have labelled.

Framework answer

Remember that although it may appear blank in a diagram, the space between structures in the skin is filled with connective tissue. Also blood vessels do not move about in the skin, they dilate and constrict altering the amount of blood passing near the surface, which is the body's contact with the environment. Diagram opposite. Functions: Epidermis, provides waterproof layer, preventing waterlogging and desiccation. Protection against bacterial infection and mechanical damage. Malpighian layer, produces new cells to replace external ones which are worn away. Dermis, cells just below Malpighian layer can develop pigment on exposure to sun, thus protecting underlying tissues from harmful effects of U.V. light. Hairs are epidermal structures but roots are deep in the dermis. In furry animals have important role in temperature control, air is trapped between, it is bad conductor of heat. Hairs raised by erector muscles, is reflex action, more air trapped, less heat passes to surface. When hairs lowered less air trapped, more heat lost. Sebaceous glands secrete sebum which oils hair, helps keep skin supple, mildly antiseptic. Sweat glands secrete sweat, comprising water, some salt and traces of urea, formed from blood. Sweat passes to surface via sweat ducts. Evaporation of water absorbs latent heat of body, cools, therefore plays part in temperature control. Nerve endings of different types sensitive to heat, cold, pressure, etc., Blood capillaries connect arterioles and veinules, blood supplies oxygen, food, removes waste matter from tissues. Capillaries also play part in temperature control by regulating amount of blood passing to body surface.

Section through human skin

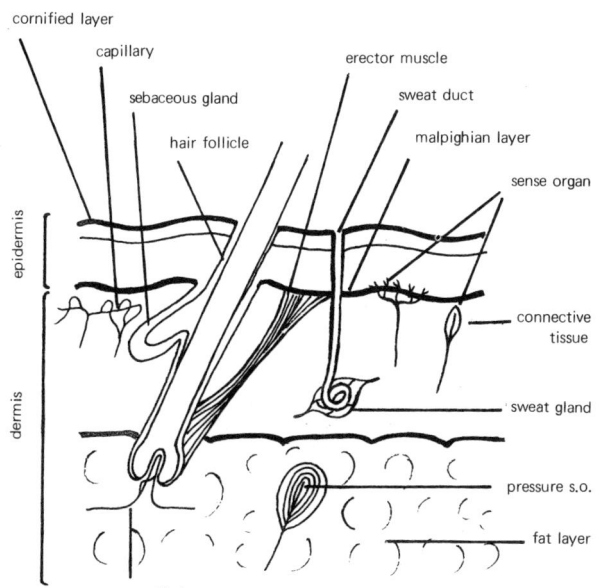

cornified layer

capillary

sebaceous gland

hair follicle

erector muscle

sweat duct

malpighian layer

sense organ

epidermis

dermis

connective tissue

sweat gland

pressure s.o.

fat layer

capillaries

Co-ordination

In a unicellular plant or animal, all the activities necessary for life take place in a single cell. Within the cell there must be organisation of the parts of the cell and all its activities. A complex multicellular organism is made up of many organs which carry out all the processes necessary for the life of each cell, plus one or more specialised jobs needed to maintain the life of the body as a whole. All the parts of the body must work together for the overall correct working of the body. The means whereby all the organs are linked to function as a 'team' is co-ordination. This is brought about (a) by the passage of electrical impulses from one part of the body to another by means of nerve cells forming a nervous system, this is found in animals only, (b) by the dispersal of soluble chemicals (e.g. hormones). These have an effect on particular parts of the body. Hormones are found in plants and animals. In animals they are secreted by endocrine glands, and often work in association with the nervous system.

Complex animals have sense organs which receive stimuli from the environment, passing information to the nervous system.

General questions

1. What is meant by the term irritability as applied to living organisms? Account for the behaviour of the iris of the eye in changing light intensities. Describe an experiment to show that the growth of a seedling shoot (or coleoptile) is affected by one-sided illumination.

2. (a) How would you determine the region of (i) perception, (ii) response of a young root? State the results you would obtain.

(b) State briefly the advantages of tropisms to plants.

3. (a) Distinguish between a reflex arc and reflex action. (b) Describe one spinal and one cranial reflex action.

4. How would you show that the skin is a sensory organ perceptive of various stimuli?

5. Make a large, clearly labelled diagram to show the structure of the human ear. Explain (a) how a sailor is able to walk across a ship's deck in rough seas, (b) how you are able to distinguish a low musical note from a high one.

6. Give a brief explanation of the following. (a) Converging lenses correct long sight, diverging lenses correct short sight. (b) Many male moths detect the female of the species even if separated by many miles. (c) After a meal, the blood sugar content does not show a dramatic increase. (d) The pair of large eyes of an insect see 'mosaic' images. (e) We are able to see near and distant objects clearly.

Framework answers

When answering questions on the eye, candidates sometimes confuse mechanisms regulating the amount of light entering the eye, and accommodation. The latter is regulation of the thickness of the lens thus focusing light from objects at different distances onto the retina.

The pupil is not a structure, it is the 'hole' in the middle of the iris. Do not write about the pupil getting larger or smaller, but of the adjustment in width of the iris affecting its diameter.

1. Irritability. The ability of an organism to react to stimuli. Any change in the environment creates a stimulus to which a response may be made.

Iris. This controls the size of the pupil, hence amount of light passing through to the retina. The muscle actions are brought about by reflexes. Retina stimulated by light, the impulses relayed to brain, then via motor nerves to iris muscles.

Pupil size and light intensity

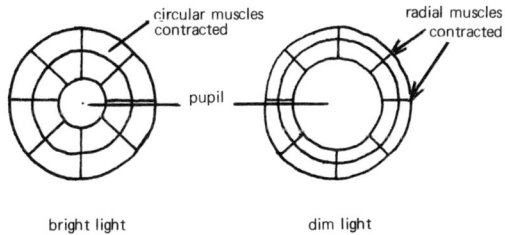

bright light   dim light

Experiment: Water soil of two pots of pea or other seedlings. Place one in box, blackened on the inside and with an aperture in one end. Direct this towards light. Place the other on a klinostat and rotate so that sides of shoot are equally illuminated. Leave to grow. Shoots of first seen to have grown and curved so that tips are pointing to aperture. Those of other, grown erect, no curvature.

2. (a) Germinate some pea seeds to give short, straight radicles. Place some in moist chamber with radicles horizontal and determine time for first signs of curvature to show. Then divide rest of seedlings into three sets, A, B, C. Mark radicles at 1 mm intervals with waterproof ink. Remove 2 mm radicle tip of each in A. Place all in moist chamber, radicles horizontal. Before curvature due to appear, treat seedlings in B as in A. Leave to grow. Radicles in B and C show curvature, those in A, do not. Removal of tip before one-sided stimulation, prevents response. Hence tip is region of perception. Examination of ink marks shows region of response is confined to a short region starting about 2 mm from tip.

(b) Positive response of roots to gravity ensures that they grow downwards and not out of soil. Negative response of the shoots, causes upward growth from germinating seeds, bringing them to light. Responses to light enables leaves to be exposed to receive maximum amount of light.

3. (a) Reflex arc consists of sensory neurone, one or more relay (internuncial) neurones and one or more motor neurones. Reflex action is an involuntary response to a stimulus, the nervous impulses involved passing along the reflex arc.

(b) A spinal reflex. Nerve impulses pass through spinal cord only, e.g. knee-jerk reflex. If patellar tendon is struck, a pull is exerted on thigh muscle, stretching it, nerve endings stimulated. Impulses via sensory nerve to cord, relayed to the motor fibre to same muscle which contracts pulling lower leg up.
Cranial reflex. Reflex arc involves brain, e.g. watering of eyes. Irritant vapours, dust, etc., stimulate nerve endings of conjunctiva, impulses pass to brain, from where impulses are sent to tear glands causing them to secrete fluid over the surface of eyes.

4. Subject of experiment should be blind-folded and asked to state what is felt when tests are applied.

Heat needle in hot water, dry and place lightly on different places on skin. Mark spots where heat is felt. Repeat with cold needle. Test sensitivity to touch by applying lightly the end of a stiff hair. Test discrimination of touch by using dividers, etc. Use both points sometimes and one at other times.

5. Make a large clearly labelled diagram.

(a) Sailor needs to maintain his balance although both he and deck are moving. Does so by means of ear which is organ of balance as well as hearing. Part of ear concerned with balance is inner ear; utriculus, sacculus, semicircular canals. In utric. and sacc. are areas with sensory processes projecting into fluid-filled cavity of inner ear. Processes embedded in gelatinous plates which contain granules. When head or body are tilted, plates pull on processes stimulating nerve endings in contact with them. Impulses travel to brain along sensory nerve cells. In ampullae of semi-circular canals are sensitive hairs stimulated by movement of fluid of inner ear, these cause impulses to pass to brain along sensory neurone. The canals are at right angles to each other and are sensitive to movement in any direction.

Movement of the ship under the sailor, is detected by sacculus and utriculus, his movement stimulates semicircular canals. Impulses from these areas travel along sensory nerve fibres to the brain. Nerve impulses from brain pass along motor nerve cells to particular muscles which contract causing the body to remain upright and move. Movement of body requires constant adjustment brought about by reflexes involving inner ear, brain and muscles.

(b) The outer and middle ear transmit sound waves to the perilymph. Running the length of cochlea is membrane made up of transverse fibres, resting on membrane are sensitive cells and suspended above these is weighted membrane. When perilymph vibrates causes some fibres to vibrate. This causes cells which stand on fibres to hit weighted membrane and emit nerve impulses. Low note causes long fibres in last part of cochlea to vibrate causing impulse to pass to brain. High note causes short fibres in first part of cochlea to vibrate, impulse to brain. Brain thus distinguishes pitch of note according to which part of cochlea stimulated.

6. (a) See opposite diagram.

(b) The antennae of moths contain the organs of smell. These are cells sensitive to very small concentrations of female odour.

(c) Presence of insulin produced by pancreas, causes excess sugar to be converted to glycogen and stored in tissues, particularly in liver.

(d) The pair of large or compound eyes of insects are made up of many small elements, ommatidia. Each has its own lens and retinal cells. The image formed is made up of a pattern of spots, one per ommatidium, hence mosaic image.

(e) See diagrams below.

Diagram of accommodation

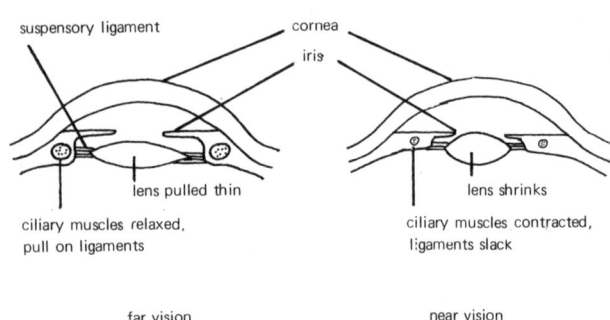

suspensory ligament

cornea

iris

lens pulled thin

ciliary muscles relaxed,
pull on ligaments

lens shrinks

ciliary muscles contracted,
ligaments slack

far vision

near vision

LONG SIGHT

b = point image
a = blurred image

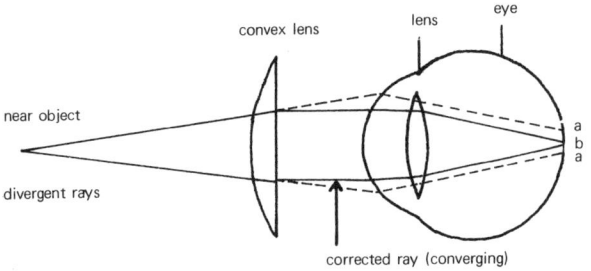

near object

divergent rays

convex lens

lens

eye

a
b
a

corrected ray (converging)

SHORT SIGHT

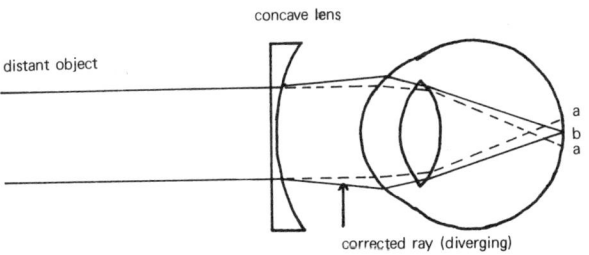

concave lens

distant object

a
b
a

corrected ray (diverging)

Movement

There is constant movement of cytoplasm in cells. Using a microscope and source of bright light, the streaming movement of the cytoplasm in cells of Elodea (Canadian pondweed) can be seen.

Some very simple plants can move from one place to another, e.g. Euglena. They are often unicellular, and move like single-celled animals using cilia (many fine protoplasmic threads), or flagella (single long thread). Euglena moves by means of a flagellum. The cilia or flagella push against the water and move the organism. The slime fungi, each of which consists of a mass of protoplasm, are capable of movement. The unicellular animal, Amoeba, also moves, but in a way distinctive to itself.

Movement of Amoeba

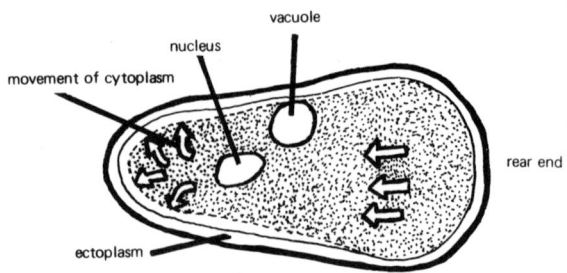

Movement is a process more typical of animals than plants. In multicellular plants it is the parts only which move, mainly by growth. The gametes of mosses and ferns move like unicellular plants and animals.

Bodies of multicellular animals are adapted to move in the different media, but movement is brought about in all cases by muscles which work as antagonistic pairs. There may be a skeleton which can be external (as in arthropods like insects) or it may be internal as in vertebrates. In biological systems as in all mechanical systems, it is considered that energy cannot be created or destroyed. Machines, including the biological one involving muscle, transform one form of energy into another. Using ATP as an intermediate, the chemical energy of sugars from food, is converted into mechanical energy for motion by muscles. In unicellular organisms, energy is transformed in the same way to bring about movement; instead of muscles, parts of the cell cause motion.

General questions

1. Explain briefly how each of the following movements is brought about, (a) the bending and stretching of Hydra, (b) the extension of the body of an earthworm, (c) the bending of an insect limb, (d) the bending and straightening of the tail of a fish.

2. (a) Using the lifting of the forearm as an example, show how movement of a limb in vertebrates is by means of a lever system.

(b) A woman puts a potted plant on a window ledge. After some days, the plant was seen to be bent towards the light. What is the difference between the woman's movement and that of the plant?

(c) By what means are bones able to move smoothly over each other?

Framework answers

1. The answer to this question will show that muscles always function in pairs working in opposition; antagonistic sets. When one of a pair contracts, the other relaxes. The principle is always the same whether the muscles pull on a skeleton or against each other, or whether the organism is pushing against air, water or earth. The essential thing to remember is that muscles can only exert force by contraction not by expansion, the volume of a muscle is constant, it is only its shape which alters.
Hydra. The body including the tentacles consists of two layers of cells. Outer ectoderm, separated by mesoglea from inner endoderm. Where cells contact the mesoglea, they are elongated to form muscle tails which contract and shorten. In ectoderm, m. tails

arranged along long axis, in endoderm m. tails aligned around the body and tentacles. Bending: On one side of tentacle ectoderm m. tails contract, endoderm m. tails relax, thus tentacle short on one side, long on the other so bends over towards shorter side.

Earthworm. Muscle cells in two layers making up body wall. The outer one encircling body, inner layer of cells lie parallel to long axis. To extend body, circular layer contracts, shortening, longitudinal layer relaxes.

Diagram of transverse section of earthworm

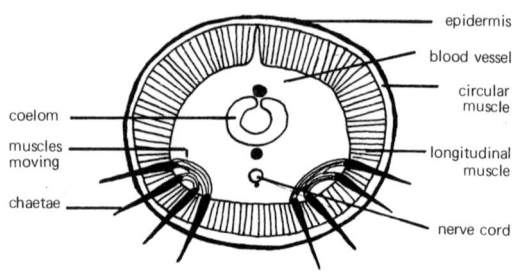

epidermis

blood vessel

circular muscle

coelom

muscles moving

longitudinal muscle

chaetae

nerve cord

Insect. See diagram. Exoskeleton of chitin. Joints, flexible unhardened parts of skeleton held together by peg and socket arrangement. Muscles attached to ingrowths of skeleton. Muscles work in pairs, flexor contracts to cause bending.

Fish. Blocks of muscle on either side of spinal cord. To bend tail, muscles on one side contract, shorten. Those on other side relax. Tail therefore shorter on one side than other so bends towards short side.

2. Bones act as levers to magnify and apply muscle contraction. To move something, an outside influence, force, (effort) is used. The product of force and the distance over which it operates

Diagram of insect limb

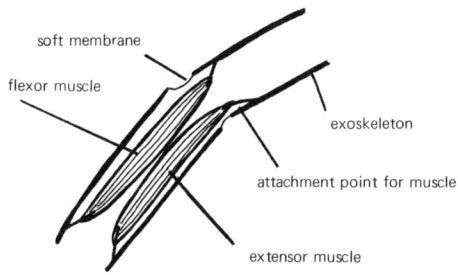

soft membrane

flexor muscle

exoskeleton

attachment point for muscle

extensor muscle

Diagram of forelimb, showing lever effect

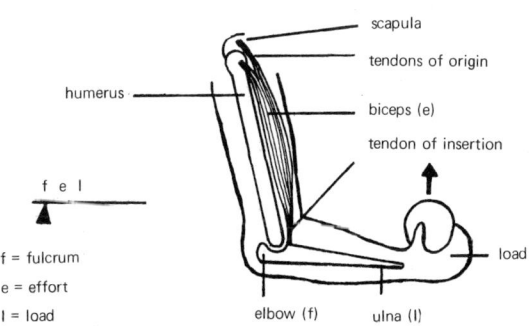

scapula

tendons of origin

humerus

biceps (e)

tendon of insertion

f e l

f = fulcrum

e = effort

l = load

load

elbow (f)   ulna (l)

is work. The point around which there is movement is called the fulcrum. The ulna acts as a lever, the elbow joint is the fulcrum, and the biceps provides the effort.

(b) Woman moves by means of muscle cells contracting, muscles shorten, pull on bones which act as levers. Energy for muscular work comes from food. Movement takes a short time, and is caused by nervous stimulation.

Plant grows towards light stimulus from one side. Light causes concentration of auxin (growth hormone) on dark side, this then makes cells increase in size so this side of plant longer than the other, so plant appears bent towards light. This takes some hours to be obvious.

(c) Surfaces of bones where they touch, covered with smooth hard cartilage, whole joint encased in capsule, within which is synovial fluid which acts as lubricant.

Support

Plants and animals need support to maintain shape and allow the transport of liquids. In addition plants have support for the leaves to expose them to light, for flowers to aid pollination, and for fruits to help dispersal. Animals need support to permit movement, and to keep internal organs in position.

Plants

Water enters cells by osmosis. This increases the cell volume, which causes the cell to push outwards against the cell wall. In mature cells, the wall will not stretch, so preventing further increase in volume. As a result the pressure inside the cell increases. Water uptake will continue until the pressure of the vacuole outwards equals the pressure exerted by the wall resisting. At this point the cell is turgid. In some plants the cells' turgidity is their only means of support. If plants lose more water than they take up, water is withdrawn from the cell vacuoles, the pressure of the vacuole decreases and the cell contents no longer push on the cell wall, the cells become flaccid. If this condition is widespread in the plant, it is seen to wilt.

Many plants also have special strengthening tissues. Some cells have thickened walls containing lignin (a woody substance). Some examples of such strengthening cells are fibres and xylem. The latter does not have so much thickening and transports water as its main function. Once lignin is deposited on cell walls, cells die. The arrangement of strengthening tissue in the stem and root is different because they have to be able to withstand different types of strain, e.g. bending or pulling. In stems the tissues are in bundles or rings forming a cylinder around the outer part of the stem, this is the arrangement most resistant to bending. Strengthening tissue forms a central strand in the root. Diagrams overleaf.

Animals

Many invertebrates have no special support tissue. The largest live in water which provides support. In soft bodied land invertebrates such as the earthworm, fluids inside the body pressing against the body wall provide support.

Stereogram of herbaceous stem

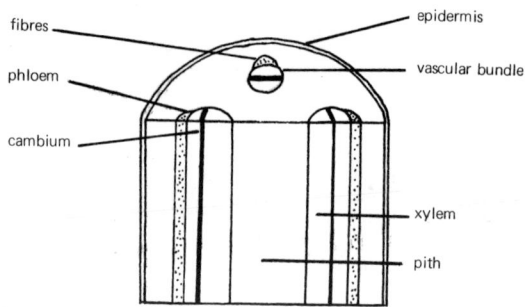

fibres — epidermis
phloem — vascular bundle
cambium — xylem
— pith

Stereogram of root

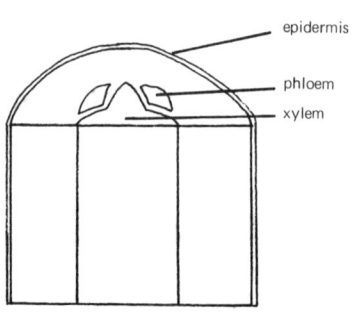

epidermis
phloem
xylem

Most land animals have a skeleton containing non-living material, but the skeleton can still grow and change shape. Arthropods have an exoskeleton which must be periodically shed (ecdysis), to allow growth. Vertebrates have an endoskeleton, as a result they can grow by continuous increase rather than in spurts as with ecdysis.

General questions

1. Describe the structure of a typical vertebra. State how you would recognise the following vertebrae: axis, thoracic and lumbar. What are the functions of the vertebral column?

2. Explain the following facts. Leaves are normally fairly rigid structures but occasionally they may lose this rigidity for a time and then recover.

3. Explain why the stems of freshly transplanted seedlings may become limp and collapse. What precautions would you take to prevent this happening?

4. (a) What are the functions of the mammalian skeleton? (b) Give the advantages of an internal skeleton over that of an external one. (c) Explain briefly the meaning of the following: axial skeleton, tendon, appendicular skeleton, clavicle, ligament.

Framework answers

1. Typical vertebra. Cylinder of bone, the centrum, surmounted by a bone arch, neural arch, forming a canal for the spinal cord. Projecting from top of neural arch, neural spine. There is transverse process on each side of centrum. Upward facing facets on prezygapophysis, downward facing on postzygapophyses, for articulation with adjoining vertebrae. Intervertebral disc covers each end of centrum.
Axis. Odontoid process on front of centrum, longitudinal neural crest rather than spine.
Thoracic. Typically, long neural spine, transverse processes on neural arch, with facet for tubercle of rib, demi-facet on each side, front and back, of centrum.
Lumbar. Large forward directed transverse processes, neural spine short. Large surface area.
Functions of vertebral column. Support, provides a series of

Diagram of typical vertebra

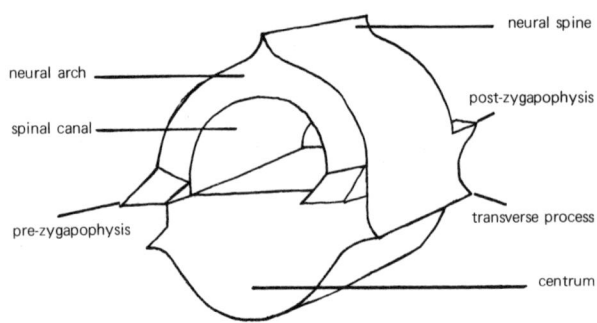

movable units with large surfaces for attachment of muscles. In sacral region provides rigid base for pelvic girdle. Thorax region provides base for pivoting ribs. Neural arches, protection of spinal cord against damage from pressure and blows. Contains red bone marrow, for making red blood cells.

2. Relative rigidity of leaves is due in part to mid-rib and network of veins, these contain vessels and tracheids with walls containing rigid lignin (wood), hence rigidity. Given an adequate supply of water the cells are fully distended with cell sap, hence rigidity. Further support may be given by presence of cells thickened, but not lignified. Loss of rigidity occurs when water is lost in transpiration faster than it can be supplied to leaf cell. Recovery, when transpiration decreases.

3. Removal of seedlings from ground destroys root hairs, thus water absorbing surface greatly reduced. Transplanted seedlings tend to lose water rapidly by evaporation, turgor lost, support lost, shoots wilt. May recover overnight if atmosphere moist and soil well watered. Full recovery unlikely until new root hairs formed.

Precautions: remove seedlings from bed carefully and with as much soil as possible, soil should be watered prior to removal. Water new site, but do not waterlog, otherwise there will be a lack of oxygen, hence respiration difficulties and slowing down of growth. Cut down transpiration by sheltering seedlings from sun and wind.

4. Functions as for Q1, also protection of brain, eyes, ears, olfactory organs. Movable lower jaw, sockets for teeth. Serves as store of calcium which may be required by pregnant females, or may be needed for bone repair.

Internal skeleton grows throughout life, a separate external protective covering also grows continuously. External skeleton shed periodically, leaves body unprotected for time. It also covers whole body surface, therefore bulky and heavy, restricts movement. Axial skeleton, skull and vertebral column. Appendicular skeleton, girdles and limb bones. Clavicle connects the sternum to scapula allowing the movement of the latter, and supporting the shoulder. Tendon attaches muscle to bone. Ligament attaches bone to bone.

Reproduction

This is the process whereby organisms increase in numbers thus ensuring the survival of their species.

Asexual reproduction On page 75 there is shown a range of organisms which can reproduce asexually.

Artificial methods of vegetative propagation

A number of plants reproduce from small pieces which eventually break away from the parent and grow into separate plants. This is vegetative reproduction. Gardeners use this property to increase the number of special plants. They take root or stem cuttings from plants such as chrysanthemums, dahlias.

Taking a stem cutting from Geranium

1. Select a shoot growing from the main stem of a geranium plant and cut it off just above a node, (point where a leaf and bud grow).

2. Remove the lower leaves, so that only 2 or 3 leaves at the top remain.

3. Fill a pot with damp soil, make a hole about 2 inches deep, put a little silver sand in the hole, plant the cutting firmly in the hole. The sand allows air to reach the cut end of the stem. A number of cuttings planted around the edge can be grown in a single pot.

4. Label the pot, giving date, name of plant and colour of flower. Put the whole pot in a polythene bag and tie. This keeps the soil moist.

5. Put the pot in a warm place. After about 4 weeks dig up the cuttings, if roots have grown transplant into separate pots.

Other methods of artificial propagation

Layering: Plants which do not usually reproduce vegetatively can be propagated by pegging the stem to the ground. Adventitious roots form from the node to grow into new plants, e.g. carnation. Budding and Grafting: E.g. fruit trees and roses. Many plants which show little natural vegetative propagation and/or somewhat

Asexual reproduction

Amoeba, binary fission

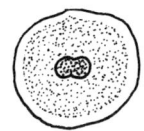 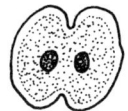 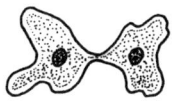

Hydra, budding

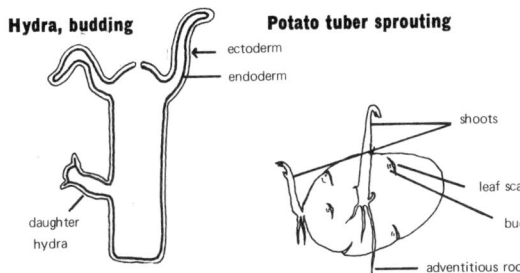

ectoderm
endoderm

daughter hydra

Potato tuber sprouting

shoots

leaf scar

bud

adventitious root

Strawberry runner

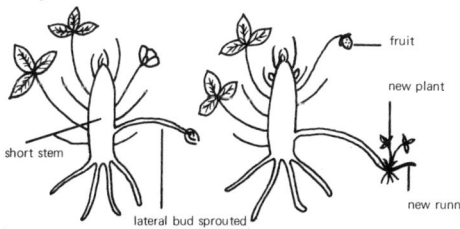

short stem

lateral bud sprouted

fruit

new plant

new runner

75

feeble growth, and which would not breed true, can be propagated by causing a bud or cutting (scion) to grow on another, related plant (stock). The essential requirement is that the cambium of scion and stock should meet. The characteristics of the scion are retained. Any shoots growing from the stock must be removed or the plant will revert to stock shoots only.

Sexual Reproduction

This involves the fusion of male and female sex cells (gametes). It is common in animals and plants. The gametes are formed by meiosis so have half the usual number of chromosomes; are haploid. Fertilisation results in a zygote with diploid number of chromosomes. Gametes may come from separate male and female parents, or as in Hydra, one individual may produce both types, although self-fertilisation does not occur. This is the fusion of gametes which were both produced by the same parent. Cross fertilisation involves fusion of gametes from two different parents. In many aquatic animals fertilisation takes place outside the body of the female; external fertilisation, e.g. frog, stickleback. Sperms have tails so can swim to the egg. Many invertebrates, most fish, amphibians, reptiles and all birds lay eggs which develop outside the body. Fertilisation in birds and mammals is always internal. The embryos of mammals always develop within the body of the mother. Flowering plants do not have mobile male gametes, as do animals and lower plants such as mosses and ferns. The transfer of pollen, containing the male gametes, to the female part of the flower, is usually by outside agent, wind or insects.

General questions

1. What is sexual reproduction? Describe sexual reproduction in Spirogyra and Mucor.

2. Give three differences between sexual and asexual reproduction. What are the disadvantages of asexual reproduction as compared to sexual reproduction?
Describe with the help of diagrams a process of sexual reproduction in Hydra.

3. (a) Describe the female reproductive system of a named mammal.
(b) Explain how fertilisation is brought about in this animal and outline the development of the offspring up to and including its birth.

4. Give an illustrated account of the male reproductive system of a named mammal. Describe the gametes produced. How do they differ from female gametes? What other functions do the ovaries and testes have, other than gamete production?

5. Give 4 differences between a floret of dandelion and a buttercup. Describe how pollination and fruit dispersal is achieved in dandelion.

6. What is meant by vegetative reproduction (propagation)? Make a list of plants which reproduce in this way by different means. Describe the process in one of the plants listed.

Framework answers

1. This is fusion of male and female gametes, resulting in a zygote which develops into a new individual.
Diagrams labelled and explained, could be used to answer this part of the question, instead of a written account.
Spirogyra: When two filaments lie beside each other, adjacent cells form a tube (conjugation canal). Cell contents round off and one moves through the canal to fuse with the contents of the other cell. Moving gamete is regarded as male, it is otherwise indistinguishable from stationery gamete. The resulting zygospore is resistant to harsh conditions. Will germinate in favourable conditions to form new filament of Spirogyra. Mucor: When hyphae of two strains of mucor grow near each other, side branches grow from the hyphae. They fuse, become cut off by cross walls, and the nuclei fuse in pairs. A resistant zygospore is formed. When conditions suitable, it germinates, forms single sporangium which releases spores. These germinate into new mycelia.

2. (a) Sexual reproduction involves production of special sexual structures. Asexual reproduction may not involve production of reproductive structures, new individuals may result from separation of parts of the body. (b) Sexual reproduction involves the fusion of 2 cells, gametes. (c) Offspring from sexual reproduction are different from the parents, offspring from asexual reproduction are genetically identical to the parents and to each other.
Disadvantages: If new individuals form from part of parent, they may not be dispersed, this results in overcrowding, e.g. strawberry.

As a result, offspring suffer from water, mineral and light shortage. As parents and offspring the same, any change in environment could result in death of all individuals. Offspring also less vigorous than individuals resulting from sexual reproduction.

Sexual reproduction in Hydra

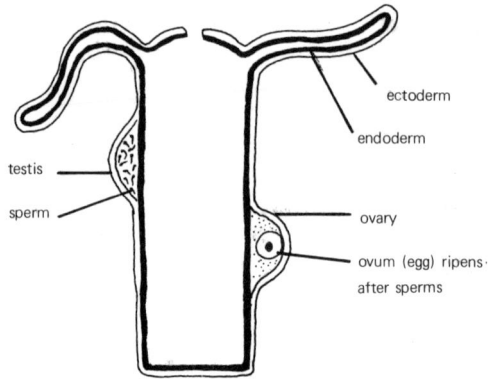

Testes arise from interstitial cells which divide to produce many sperms, causes ectoderm to bulge near tentacles. Ovaries also form interstitial cells which divide, causing bulge below ectoderm near base. One cell grows at expense of others to form egg. Ectoderm splits first, sperms released, swim towards nearest Hydra with ripe egg, attracted by chemicals. Ectoderm then splits to expose egg. After fertilisation, zygote with protective cover falls off. New Hydra emerges in spring, grows from original cell by division and enlargement.

3. E.g. man. Diagram and labels: ovary, fallopian funnel, fallopian tube (oviduct), uterus, vagina, vulva.

Ovaries: paired, pink, blistered appearance, lying just below (posterior to) kidneys, more or less enveloped by fallopian funnels. These have ciliated edges to sweep ovum into tube. This narrow tube leads to muscular walled uterus, which is well supplied with blood vessels. Uterus is pear-shaped, there is narrow aperture in cervix leading to vagina, a wider dilatable tube connected to exterior by vulva, an opening with fleshy lips.

Fertilisation. Seminal fluid containing sperms, travels along penis which is inserted into vagina. Sperms swim, helped by movements of uterus, to fallopian tubes. Ovum discharged from ovary, swept into tube by ciliary action of funnel cells. Meeting sperms, one penetrates the ovum, sperm and ovum nuclei fuse, fertilisation taking place.

Development. Fertilised egg divides immediately to form ball of cells, moved to uterus. Outer part forms foetal membranes, inner mass of cells forms embryo. Whole implanted in uterine wall. Rapid development of amnion to form liquid filled cavity around embryo. Gut begins to form and blood system. Outgrowths of allantois from hind gut grow out to form part of placenta embedded in uterine wall. Blood of embryo brought into close contact with maternal blood. (Note. Blood of mother does not flow into the embryo or vice-versa.) Embryo grows, tissues and organs develop as it receives food, oxygen via umbilical cord from placenta. Prior to birth, rhythmic contractions of uterine muscles start, become more frequent, pressing on baby and containing sac. Aperture of cervix widens, wall of sac ruptures, fluid escapes, baby forced usually head first, into vagina and out. Umbilical cord tied and cut, breathing starts. Placenta expelled as afterbirth.

4. Diagram and labels. Scrotum, testis, epididymis, seminal vesicle, prostate gland, urethra, penis.

Testes. Paired, in scrotum, consist of coiled tubules in which sperms are produced, and connective tissue containing cells producing male sex hormones. Tubules lead to large collecting tube, epididymis inside of testis, to vas deferens, which passes into abdominal cavity. Vas deferens coils around ureter, leads to urethra. Seminal vesicles and prostate provide fluid stimulating sperms to activity. Sperm. Minute, head is nucleus with thin cover of cytoplasm, middle piece with mitochondria. Long tail with axial filament. Differs from ovum in size, shape, motility. Half contain Y chromosome.

Both organs produce hormones, female and male sex hormones respectively. These control sexual maturity and in the female menstrual cycle and changes associated with pregnancy.

5. (a) Floret, one ovary and ovule, buttercup, many. (b) 5 anthers in floret, many in buttercup. (c) Petals fused in floret, separate in buttercup. (d) Calyx pappus of hairs in floret, 5 green sepals in buttercup.

Pollination. Small insects crawling over florets pick up pollen, visit other flowers, pollen brushed off onto stigmas. If cross pollination does not occur, stigma may curl back and touch the anthers picking up pollen.

Dispersal. Pappus of hairs well developed, increase surface area, so fruit caught in air currents, floats down slowly so carried considerable distance from parent.

6. Vegetative reproduction. Formation of new individuals from buds developed on parent plant. New plants take root and eventually lose connection with parent.

Daffodil (tulip, snowdrop) by bulbs. Crocus (gladiolus) by corms. Potato (Jerusalem artichoke) by stem tubers. Dahlia (lesser celandine) by bulbils.

Potato. Lateral buds on stems below ground level for shoots, these stay below ground. Food made in leaves passes through phloem to these stem tips, diffuses into parenchymatous cells mainly pith. Glucose converted to starch grains, protein stored mainly under thin corky skin. Cells increase in size and number to accommodate food. Parent plant dies in late summer, leaving a number of tubers which perennate. Each can produce one or more new plants the following year, hence increase in numbers.

Genetics

The material of inheritance, genes, are carried on chromosomes in the nucleus. The chromosomes are clearly visible during cell division. They can be seen in stained sections of the area just behind the root tip on onion. Growth takes place here and the cells are dividing by mitosis. Slides of cells from locust testes also show chromosomes, here the cells are dividing by meiosis to form sperms. No two organisms or part of the same organism are seen to be completely identical, thus variation is a characteristic of living things. All living things are grouped into kingdoms, phyla, classes, etc., in such a way that we can see relationships. E.g. tigers, lions, domestic cat are all in the same family, but are obviously very different. They do not gradually merge into one another. This is called **discontinuous variation**. If a large number of individuals of the same species are examined, it is found that qualities like height, weight and colour, etc., are never all the same in any two individuals. If you measured the height of all the people in your class, you would find that it varies from very small to very large. There are examples of both extremes, with every gradation in between. This is **continuous variation**.

The original work on genetics was carried out by Abbé Mendel. He found that if he selected pea plants with contrasting characters, e.g., tallness and shortness, and crossed them, all the offspring of this 1st generation (F_1) were tall. He then allowed these F_1 individuals to self pollinate, the resulting generation (F_2), were in the ratio 3 tall: 1 short. There must be a pair of genes (alleles) responsible for height in the pea plant, one carried on each of a pair of chromosomes. The gene for height can be for tallness or shortness. When gametes are formed, they have a haploid number of chromosomes, so each has only one gene for each character. T represents the gene for tallness, t represents the gene for shortness.

$$\text{Parents} \quad \underset{\text{(tall)}}{\text{TT}} \quad \times \quad \underset{\text{(short)}}{\text{tt}}$$

Gametes T T t t   each has only one of a
pair of alleles.

Gametes fuse in fertilisation, zygotes now have diploid number of chromosomes, therefore both genes for height.

	T	T
t	Tt	Tt
t	Tt	Tt

all F_1 offspring tall,
though have genes for
tallness and shortness.

F_1 plants are heterozygous; contain genes for tallness and shortness. As they are all tall, this is the dominant gene, the gene for shortness is recessive.

When F_1 self pollinated

Parent genotype Tt $\times$ Tt
gametes   T   t   T   t

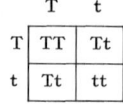

	T	t
T	TT	Tt
t	Tt	tt

Results. **3** tall: 1 short.

Backcross

In order to find out which genes may be present in the heterozygous individual of F_1 generation, a backcross is carried out. This is crossing the F_1 generation with the recessive homozygous parent.

Artificial selection

Man has cultivated crops and kept domestic stock for thousands of years. He has selected certain features, e.g. high yield in crops, high milk production in cattle, and by artificial selection has bred crops and cattle which have these characteristics. The cattle with high milk production, and those whose ancestors and relations also gave high yields, were crossed. By means of similar artificial selection, high yield crops were bred.

General questions

Genetics questions as well as those involving simple calculations are among the few biological questions that can easily earn you almost full marks. Provided that is, you understand the basic principles involved in crossing individuals having a pair of alleles for contrasting characters. Remember that parents always have two genes, but gametes only contain one, and that fertilisation is random. You must also understand the meanings of the terms used in genetics and must use these terms wherever possible.

1. When Mendel crossed pure ſbreeding pea plants that were tall with plants that were short, all the F_1 generation were tall. He allowed these to self pollinate and he obtained 787 tall plants and 277 short ones. Explain these results.
How would you establish the genotype of one of the tall plants of the F_2 generation?

2. Draw a single diagram to illustrate a cross between a male fly heterozygous for grey body colour, with a female of black body colour. Body colour is controlled by a single pair of alleles, grey colour being dominant. Use the symbol G to represent the dominant allele and g to represent the recessive allele.

(a) What proportion of the offspring would you expect to be homozygous grey?

(b) What result would you expect if a black male was crossed with a grey heterozygous female?

3. Mrs Brown's father had been born with a sixth finger, but neither Mrs Brown nor her husband had this abnormal characteristic. They had two children, Pat and Jim, only Jim had this extra finger. Give a simple genetical explanation of its inheritance. Pat eventually married and her husband had normal hands.

What are the possible genotypes of Pat and her husband?

Framework answer

1. The gene for tallness dominant, gene for shortness recessive. One set of parent plants homozygous for tallness, 2 alleles for tallness (TT), other set homozygous for shortness, 2 alleles for shortness (tt). Diagram below represents cross between one set of parents.

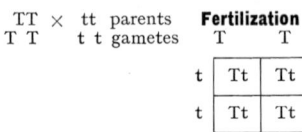

Phenotype (appearance) of all F_1 generation tall, genotype Tt heterozygous.

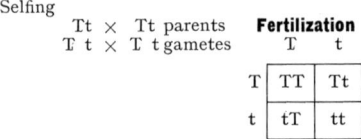

F_2 generation 3 tall: 1 short.
787 tall, 277 short, the proportions are approximately 3:1. In order to establish the genotype of one of the tall plants of F_2 generation, a backcross must be carried out. This means to cross the tall F_2 plant with the homozygous small parent. Tall plant could have genotype TT or Tt. Results of cross establish which genotype.

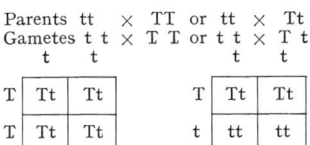

Parents tt × TT or tt × Tt
Gametes t t × T T or t t × T t
           t   t

	T	T
T	Tt	Tt
T	Tt	Tt

	T	T
T	Tt	Tt
t	tt	tt

If F_2 homozygous, all off-spring tall.

If F_2 heterozygous, offspring half tall, half short.

2.

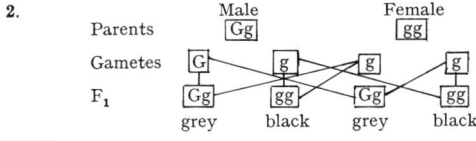

Parents   Male Gg   Female gg

Gametes   G   g   g   g

F_1 Gg   gg   Gg   gg
    grey   black   grey   black

(a) None

(b)

		Male	
		g	g
Female	G	Gg	Gg
	g	gg	gg

Results. Half black
         Half grey

3. Gene for 6th finger is recessive, as Mrs. Brown did not show the extra finger, but she carried the gene inherited from her father and passed the abnormality on to her child. Jim must have had homozygous alleles for a 6th finger, so his father must also have had the recessive gene which was 'covered' by the dominant gene for normal hands. Let F represent the gene for normal hands, f represents the gene (allele) for 6th finger.

       Mrs Brown
       F   f

		F	f
Mr Brown	F	FF	Ff
	f	Ff	ff

Jim's genotype ff homozygous for extra finger

Possible genotypes of Pat and her husband, Ff (heterozygous normal), FF (homozygous normal).

Growth and Development

Growth is an increase in size, brought about by cell division and by cell enlargement. New protoplasm is made from food. The plant cell may increase in size by enlargement of the vacuole, due to water uptake. Increase in weight usually accompanies growth which can be studied by measuring changes in size and weight. Many green plants, including all those that flower, grow in certain regions only; these are meristems. Increase in length occurs at shoot tips, buds which are compressed shoots, and just behind root tips. Growth in thickness is brought about by another meristem (cambium) forming a ring round stem and root. Growth in animals occurs throughout the whole body. Many plants grow continuously throughout life, others such as annuals reach a certain maximum size then cease growing. In plants growth is seasonal. Many invertebrates grow throughout life, as in plants the rate decreases with age. Land animals such as mammals and birds grow to a certain maximum size, then growth ceases. Arthropods with a hard exoskeleton can only grow if ecdysis occurs, when growth takes place in a rapid spurt. Vertebrates grow slowly and continuously until they reach a certain maximum.

Factors affecting growth. Growth will be limited by lack of food in animals and the substances needed to make food in plants. Increase in temperature, up to a certain maximum, causes increase in growth. Greater light intensity causes increase in photosynthesis in green plants. More food is available so more growth occurs. Extreme acid or alkaline conditions have an adverse effect on growth. Growth in plants and animals is controlled by hormones.

Development is all the changes involved in producing a mature organism. From a fertilised egg or zygote, develops the complex structures that make up an adult organism. Development is controlled by genes inherited from parents, which are in the nucleus. Chemicals, including hormones, in the cytoplasm also have an effect on development, as does environment. Exactly how different tissues form from one cell is not fully understood. The process whereby different tissues are formed is differentiation.

General Plant Topics

Perennation is the process whereby plants remain alive during the winter, surviving for many seasons. Such plants are perennials. They may survive with reduced foliage above ground, e.g. grass, or they may have **an** underground storage organ, e.g. crocus corm. These storage organs may also be a means of asexual reproduction. Woody perennials are trees and shrubs.

Fruits and seeds

After fertilisation, petals, stamens, style, stigma fall off. The ovary becomes the fruit, food for its development comes from the leaves of the plant. The ovules become the seeds, each of which contains an embryo plant. This consists of a tiny root, (radicle) shoot, (plumule), one or more leaves (cotyledons). These may contain stored food. The integuments of the ovules become the hardened testas. Water is withdrawn from the seed as in this dry condition it is better able to withstand adverse conditions.

Seed dispersal. This is necessary (a) to prevent overcrowding which would result in shortage of water, salts and light, (b) to permit colonisation of new habitats. There are four main agencies for dispersal; wind, animals, water and self dispersal.

Seed structure. There is a tiny pore (micropyle) in the testa, there is a food store, usually starch, and other substances such as oil. The food store is always insoluble, and may be in cotyledons or in a separate endosperm. Monocotyledons have one cotyledon, dicotyledons have two cotyledons.

Germination. This is development of the embryo after a dormant period, the length of which is often determined by external factors, e.g. frost. Water, warmth and oxygen are all necessary for germination to occur. Some seeds also need light, e.g. onion. Germination begins with water absorption through the micropyle. Seed swells, respiration increases, insoluble food changed by enzymes into soluble food which is transported to the radicle and plumule in hypogeal forms. Germination is of two types. Hypogeal, cotyledons remain below ground either as a food store or absorbing food stored in the endosperm. Epigeal, the cotyledons form the first green leaves and are pushed above ground.

Germination

Hypogeal Germination — Pea

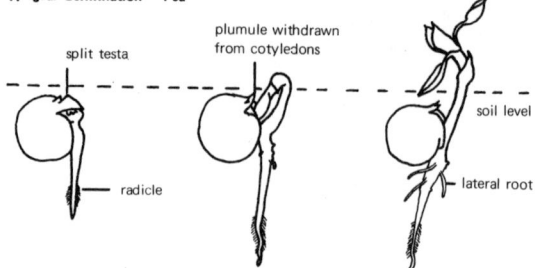

split testa

plumule withdrawn
from cotyledons

soil level

radicle

lateral root

epigeal germination — sunflower

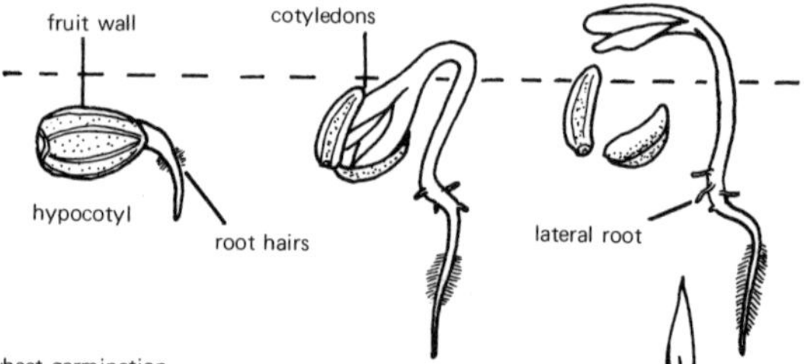

fruit wall

cotyledons

hypocotyl

root hairs

lateral root

wheat germination

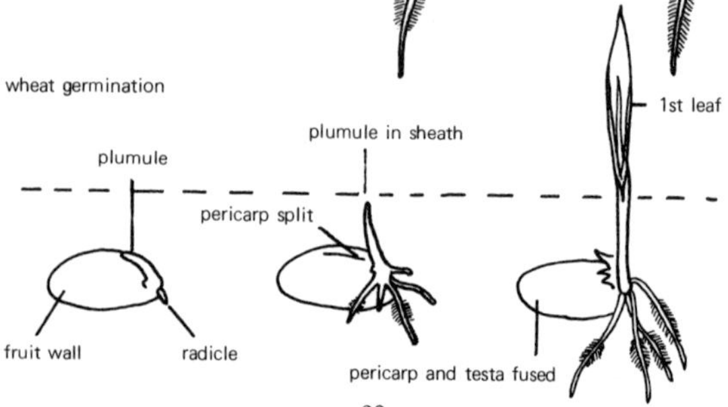

1st leaf

plumule

plumule in sheath

pericarp split

fruit wall

radicle

pericarp and testa fused

88

Leaf fall. Deciduous trees shed all their leaves in the autumn, coniferous trees shed a few all the year. Leaf fall in sycamore is shown below. At the base of the leaf stalk is a special layer of cells, the abscission layer. In autumn, the cell contents of deciduous leaves are chemically changed, become soluble and are transported back into the tree to be stored. This is why leaves change colour. The inner cells of the abscission layer become corky and the veins become blocked, so the leaf is deprived of water. The cells beyond the corky layer become jelly-like and the leaf falls off. The corky layer protects the tree against infection by bacteria and viruses.

Leaf fall

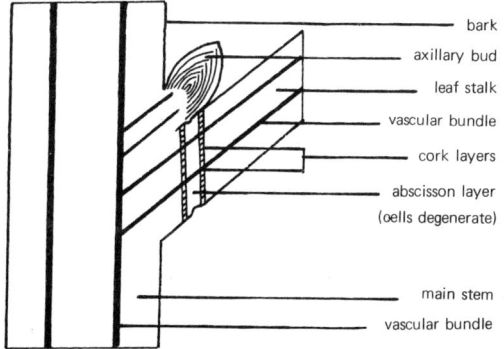

bark

axillary bud

leaf stalk

vascular bundle

cork layers

abscisson layer
(oells degenerate)

main stem

vascular bundle

Horse chestnut twig

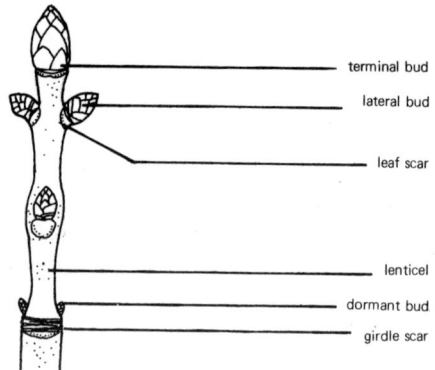

terminal bud

lateral bud

leaf scar

lenticel

dormant bud

girdle scar

4 year old woody stem

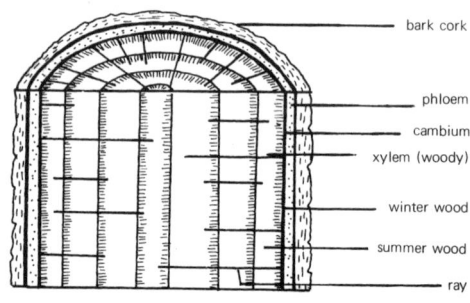

bark cork

phloem

cambium

xylem (woody)

winter wood

summer wood

ray

Soil

The study of the components of soil can be carried out entirely by experiment. Experiments carried out should include those which find out how much air, water and humus make up a soil sample. Permeability of clay and sand to air and water should also have been studied practically. You should know at least one method for demonstrating the presence of micro-organisms in the soil. Remember to use sterile soil samples as controls.

Question

(a) Explain two ways in which bacteria improve soil fertility for plants. (b) What is the importance of the insoluble organic compounds in soil to bacteria? (c) How is nitrogen released into the atmosphere?

Framework answer

Many candidates answer all questions on soil bacteria by a diagram of the nitrogen cycle. Often this is the only attempt made to answer the question. Make sure your answer is relevant and do not attempt questions on soil bacteria unless you understand the topic well.

(a) Organic matter decomposed by feeding bacteria, forming humus. This improves soil by retaining water and air in soil, prevents waterlogging.
Nitrogen fixing bacteria fix atmospheric nitrogen, build it in to protoplasm. Bacteria die and decay, nitrification, yields nitrites and nitrates.

Or:

Decay bacteria release ammonium compounds, these converted by Nitrosomonas (nitrifying) bacteria into nitrites, converted by Nitrobacter into nitrates.

(b) Food for bacteria, used to give energy for growth and reproduction.

(c) Denitrifying bacteria in badly aerated soil respire by breaking nitrates to oxygen and free nitrogen.

Ecology

Ecology is the study of living things in their natural surroundings. All the organisms living in a locality have an effect on each other and on their environment. Plants and animals live as related communities in a range of habitats, e.g. seashore, ponds, woodlands. They show differing degrees of adaptation. Any habitat is made up of 4 components. 1. Non-living substances, soil particles, water and substances dissolved in it. 2. Organisms making food from non-living materials, e.g. green plants. 3. Organisms feeding on other living things, e.g. animals. 4. Decomposing organisms, e.g. fungi and bacteria. Soil components and climatic conditions will determine which plants grow in a locality. Animals feed on the plants and are in turn eaten by other animals. Thus the organisms in a community are linked by **food chains.** When animals and plants die their bodies decompose so the mineral salts are returned to the soil. If food chains are studied, it will be found that at the bottom of the chain there are enormous numbers of herbivores, these are eaten by small carnivores, in smaller numbers. At the top of the chain are a few large carnivores. This comparison of quantities of animal matter involved at the different stages of the food chain, can be represented as a **pyramid of numbers.** There are two types of ecological study; (a) study of a single species, including structure, distribution, behaviour etc., (b) study of plants and animals which occur together in a community. Data should be written up neatly in a field notebook. Collect as few specimens as possible and only very common species. The following data would be required if studying a community: 1. Description of habitat. 2. Date and time of observations. 3. Physical factors, temperature, pH, humidity, wind etc. 4. List of species found. 5. Size of sampling area, method, i.e. transect, quadrat. 6. Numbers of species found. 7. Behaviour of animals and exactly where found. Fresh situations are immediately colonised by organisms, these will be followed by others, **succession**. When this ends and a stable community established, a **climax** community exists. Man's activities provide fresh situations, and he prevents the climax from becoming established (e.g. by keeping grazing animals etc.); causes a **sub climax** situation.

Suggestions for Further Practical Work

The following suggestions might provide some ideas for extending the normal practical work, or they could be used as the basis of C.S.E. projects.

1. The study of a habitat. Many areas are worth investigation if you have easy access to the countryside, e.g. woodland, chalk hills, heath, rocky shores and sand dunes. However, this is not always possible. Your own garden can also provide many opportunities for investigations. As well as the plants in the garden, there will be vast numbers of small, invertebrate animals. These will be found in the soil, under stones, on leaves, under the bark of trees and in the compost heap.

(a) **Earthworms.** Investigations could be carried out into their food preferences, materials used for plugging burrows, amounts of leaf litter drawn into the soil and amounts of soil raised to the surface as worm-casts.

(b) **Snails.** It is possible to investigate their movements by marking the snail's shell with paint and observing their behaviour over a period of time. Records could also be kept of the distances travelled. Experiments could be carried out to determine their food preferences and the conditions which will bring about aestivation. Similar observations can be made with slugs.

(c) **Ants.** A simple ant-nest can be constructed with soil between two pieces of glass or perspex. These can be bound together with tape leaving a small entrance hole. If this is stood on a brick standing in a dish of water they will not be able to escape. One queen, about 50 worker ants and about 20 pupae are needed, from the same ant-nest. These are placed on the surface of the ant-nest, which should be covered to prevent entry of light. The movements of the ants into the nest can then be observed. The ants can be fed on sugar or honey placed on the lid of the nest. Once the nest is established many additional investigations are possible.

(d) **Caterpillars and butterflies.** Caterpillars are easily kept providing their food plant is known. It should be remembered that each type of caterpillar has certain plants that it will eat and will not survive without the correct type. Experiments could be conducted to determine the range of plants they will feed on, the effects of the background colour during the larval stage on the colour of the pupae, the methods of spinning the cocoon and, with some, mating and egg-laying in the adult.

(e) **Spiders.** These can either be observed in their natural surroundings or in captivity. Web-building is one of the most interesting aspects of spider behaviour. Observations could be made on the methods of web construction, the range of food caught in the web and the spider's behaviour should the web become damaged.

(f) **Woodlice.** These can be kept in a small container with a slice of potato. As with snails, they can be marked with paint and their movements in the garden can be followed. Their responses to light, dark, moist and dry conditions can be studied in a simple choice chamber.

Similar investigations can be carried out with other invertebrates.

2. **An aquatic habitat.** Rivers, streams, lakes and ponds are interesting places for study. Where a suitable site is not available, an old container filled with water and left in the open will soon be colonised by many plants and animals.

(a) **Microscopic organisms.** There are many possible investigations, including their distribution in different types of water, the variation of types and numbers with depth and their seasonal variation.

(b) **Hydra.** Feeding can be observed under the microscope. Their method of feeding and reproduction could make an interesting study.

(c) **Flatworms.** These are very common in most rivers and streams. Their food preferences, movements in currents of water and their regeneration when cut into small pieces could be investigated.

(d) **Caddis larvae.** These larvae build themselves tubes from leaves, twigs or stones. Certain types, especially those with leafy tubes, will rebuild their tubes in captivity. If various materials are provided, e.g. pieces of coloured paper, plastic, beads, cloth, flower petals, etc., they will use these for their tube instead of leaves. An interesting study could be made of the preferences shown for particular materials.

(e) **Pond snails.** Studies could be made of their movements, mating and development of young.

Complete List of
Key Facts Educational Aids

KEY FACTS COURSE COMPANION BOOKS

Physics
Chemistry
Biology
Modern Mathematics
French
Geography

English
Economics
Additional Mathematics
Arithmetic and Trigonometry
Algebra
Geometry

KEY FACTS CARDS

English Language and
   Examination Essay
English Comprehension and Precis
Geography
History
French
Biology
Chemistry
Physics
Modern Mathematics
Elementary Mathematics
Additional Mathematics

Arithmetic and Trigonometry
General Science
Economics
Geography—Regional
Algebra
Geometry
Technical Drawing
Latin
German
Macbeth
Julius Caesar
New Testament

KEY FACTS 'A' LEVEL BOOKS

Pure Mathematics
Physics

Biology
Chemistry

Published in Great Britain by
Intercontinental Book Productions
in conjunction with and available from
SEYMOUR PRESS LIMITED, 334 Brixton Road,
London SW9